SOLAR ENERGY – A BOON

COMPLETE GUIDE TO BUILD A SOLAR PLANT FOR OWN USE.

MANOJ M KALKE

Dedicated to all MSMEs in India

who always struggle for

Excellence

&

many times even for Survival

but

they never give up !!!

Contents

Preface — vii

Acknowledgements — ix

TABLE OF CONTENTS — xi

1. ENERGY - CURRENT SCENARIO — 1

2. WHAT IS SOLAR ENERGY ? — 5

3. BASIC ELECTRICITY — 9

4. ELECTRICAL SAFETY — 18

5. TYPES OF SOLAR SYSTEMS — 21

6. COMPONENTS OF THE SOLAR SYSTEMS — 29

7. DESIGN OF THE SOLAR ENERGY SYSTEMS — 43

8. INSTALLATION OF THE SOLAR ENERGY SYSTEM — 69

9. COST & MAINTENANCE OF SOLAR ENERGY SYSTEM — 79

10. INNOVATIVE PRODUCTS — 82

11. FUTURE — 87

12. SCHEMES FOR SUBSIDY FOR SOLAR ENERGY — 89

13. REFERENCES — 102

Preface

PREFACE

In our Sanatan Hindu religion, the sun is considered a deity. We believe that SUNLIGHT is the cause of life on earth. And it is also true that this energy source is eternal. Therefore, a book that provides information about solar energy in simple language is the need of the hour. The author has given detailed information on solar energy, its national and international importance, and current hot topics like 'Climate Change' in the very first chapter. If the reader reads the first chapter carefully, I think that this introduction of mine was not intended, because I am not an expert in the subject of Solar Energy'. On the contrary, after reading this book, I got detailed information about 'Solar Energy'.

The subject has been arranged very thoughtfully here. The author has explained the stages of 'Solar Energy to generation and use of electrical energy' in simple language, giving appropriate synonyms to all the dimensions and terms, and with the help of diagrams. Instead of just providing descriptive information, the subject has been made clearer by giving examples like a workshop for those who are directly working. If the book is read carefully from beginning till end, the reader will feel like understanding a topic in detail.

As the author says, this book will introduce a new energy source to people who are interested in science. It will be useful for those who do technical work, such as electricians, wiremen and maintenance mechanics. Moreover, the information on how to use solar energy in household appliances and how to do its preliminary planning will also be very useful for ordinary citizens.

The author, Mr. Manoj Kalke, must have considered my plan to write the introduction to the book as a well-wisher and a representative of the potential readership of his book. Otherwise, I admit that both the subjects 'Solar Energy or Literary Quality ' are not mine. It can be said that he has served Maharashtra in a way by writing this somewhat complicated subject in simple language.

His writings show a deep study of the subject, scientific attitude and vast work experience. I appreciate, congratulate and wish him all the best for the efforts taken by Mr. Kalke for the book on an innovative and remarkable subject.

May 'Gramdaivat Kanakaditya' inspire him to undertake such new initiatives in the future and may he gain fame as a successful entrepreneur.

Mr. Prakash Gune

Founder and Managing Director
Pragati Electricals Pvt. Ltd., Thane

Acknowledgements

I am thankful to my friends Shri. Ravindra Sawant and Shri. Mahesh Hejmadi to always encourage and support me in all ways in whatever I do. Their word of appreciation for my work has been a source of energy for me.

I thank my nephew, Durvesh Kalke, a researcher scholar in Electrical Engineering field from Indian Institute of Science, Bangalore, who helped me with his expertise in the field.

Last but not the least, I thank, NotionPress Team for providing so easy to use online platform to publish a book and superb customer support they provide.

Table Of Contents

Table of Contents

PREFACE

1) Energy - Current Scenario

2) What is Solar Energy?

3) Basic Electricity

4) Electrical Safety

5) Types Solar Energy Systems

6) Components of Solar Energy Systems

7) Design of Solar Energy Systems

8) Installation of Solar Energy Systems

9) Cost & Maintenance of Solar Energy Systems

10) Innovative Products

11) Future

12) Schemes for Subsidy for Solar Power

REFERENCES

CHAPTER ONE

ENERGY – CURRENT SCENARIO

Renewable Energy, Green Energy, Ecofriendly products, environment and nature and such words have now become "Buzzwords". Such words have started to get a place of honor in reading everyday writings, in fact, if we use such words in our conversations, people look at us with respect. Still, it is a good thing that people have now started talking about the topic of environment and nature. In this book, we are going to learn about one of the many latest energy sources, Solar Energy.

Renewable energy is energy that does not decay or decrease after use, but is able to provide new energy again. Solar, Wind, Hydroelectric, Biogas and Biomass are all renewable energy sources. In this book, we will only learn about solar energy. In daily use, we use the words Power and Energy in the same sense or interchangeably; but physics differentiates between the two. Physics says that energy is the ability to apply power. Anyway, in this book, we will use the word energy in the sense of electrical energy. Also, since this information is technical in nature and though many English words are more familiar to us, we will use examples in daily life for better understanding of the subject.

Today, the energy scenario in our country is not as good as it should be. We are a developing nation and our economy has also become dynamic. For sustainable development and economic progress, it has become necessary to have uninterrupted electricity supply to factories, commercial establishments and households of all levels of society. This is called energy security and it has become as important as food security. The issue of energy security can be solved through government policies and far-sighted strategic decisions. Sustainable development and economic progress unfortunately have a direct impact on the environment. It is also related to climate change. The negative impact of progress and industrial development on the environment can be offset to a large extent by using solar energy. Ever increasing power requirement can be met to some extent by Solar Energy in our country.

Solar Energy – Current Scenario

Interesting Statistics – We require huge POWER to become developed economy and to sustain the development

Country	Per Capita Electricity Units Consumption	Economy / GDP 2025 $ Tr
US	16000	30
China	14000	19.6
Germany	13500	5
Japan	13000	4.4
India	1000	4.3

Huge Requirement of power in the Country

The statistics of our country regarding energy consumption are worth studying. Today, we still mainly use fossil fuels for energy. We meet our needs using coal - 40%, oil - 22%, gas - 8% and the remaining 30% from other sources. Out of this other 30% of energy, we use only 1% of solar energy. The potential to generate solar energy in our country is huge! In comparison, despite our geographical location being favorable for solar energy, we generate very little solar energy. India is the sixth largest energy or electricity consumer in the world. If we consider the world, 80% of the population lives in underdeveloped and developing countries and these 80% of people use only 30% of the electricity generated in the world. There is a huge gap between electricity production and electricity consumption in our country. Our total electricity production is 360 GW, out of which only

300 GW is available for consumption. This difference is due to maintenance, line faults, power theft, and the time difference between power generation and demand. The demand for electricity is increasing day by day, and solar energy is a better alternative to fossil fuel-based electricity.

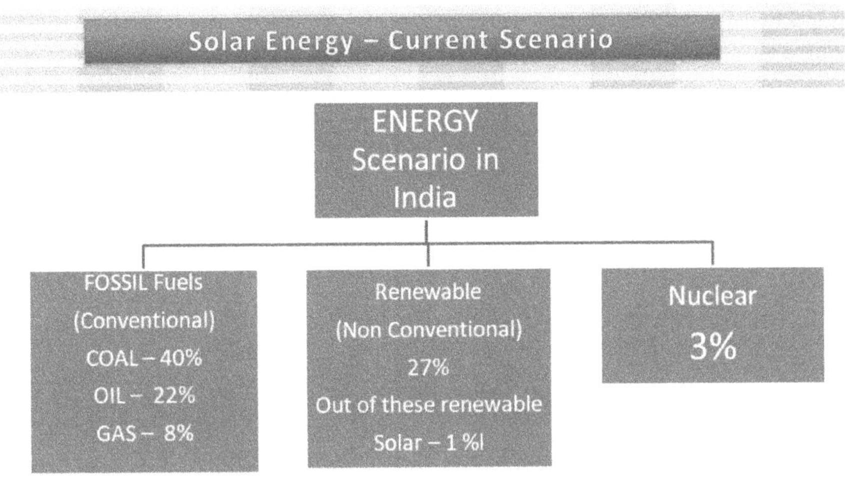

Enter Caption

Where electricity is available in abundance, the area develops rapidly. Where there is development, the demand for electricity increases. The government divides the area into urban, semi-urban and rural according to population. The availability of electricity in rural arcas is very low and due to this, the pace of development is also very slow. Rural areas are deprived of amenities due to unavailability of continuous power. Abundant availability of electricity can lead to sustainable development of rural areas and can improve the standard of living of the people. In India, the government started the rural electrification program in 1950. Due to remote areas and far-flung houses, electricity could not reach every household till date. The task of supplying electricity to households through the national grid was becoming expensive and these government schemes did not have much impact until 2016. Solar energy can be an effective solution to this problem. It is easier and less expensive to generate electricity for that village or villages and supply electricity to households. Installing solar panels on the roof of each house and generating electricity for that house is also a good option. In rural areas, every house, establishment or factory can become self-sufficient in terms of electricity by using solar energy.

Solar energy is now not only an affordable option for domestic use, hospitals, schools, agriculture, communication, or any other use, but also it can definitely be a viable and healthy option. If training is provided in the rural areas of the country on how to generate energy as well as how to use it, the need for energy can be reduced. For example, many people install pumps in cities and villages without knowing how much capacity of a pump can meet their water needs. For example, if my daily water needs are met by a 2 HP pump, many people install a 5 HP pump thinking, "If I'm going to install one, I'll install a bigger pump." This situation is very obvious. In this situation, we increase our electricity bill unnecessarily. If such training is provided in rural areas, identifying our electricity needs and generating enough electricity, the need for electricity can also be reduced and if solar energy is generated for our consumption, the rural areas will become self-sufficient and their development can be achieved quickly and sustainably.

The purpose of this book is not just to collect information, but by using this information, non-technical people, whether they are in urban, semi-urban or even rural areas, can become self-sufficient in terms of energy by generating solar energy. This is an attempt to present this information in simple and useful words for people without a technical background. The knowledge of generating solar energy can also become a means of business or self-employment for many people. There are also some villages in the country where access to traditional energy, i.e. electricity, will be difficult even for the next few years. The information in this book has been compiled with the aim of being able to meet our electricity needs by putting it into practical use, so many demonstrations and pictures have been used abundantly. Solar energy revolution is bound to happen, let us participate in it on time and become self-sufficient in energy.

CHAPTER TWO

WHAT IS SOLAR ENERGY ?

Solar energy is available in huge quantities in our country. India generally has 300 days and 6 hours of clear or strong sunlight every day. We want to use this abundant sunlight to generate electricity. Electricity generated from sunlight is called solar energy. We can use sunlight in two ways 1. To generate thermal energy and 2. To generate electrical energy i.e. electricity. In this book, we are going to learn the information and techniques to convert sunlight into electrical energy only.

Construction

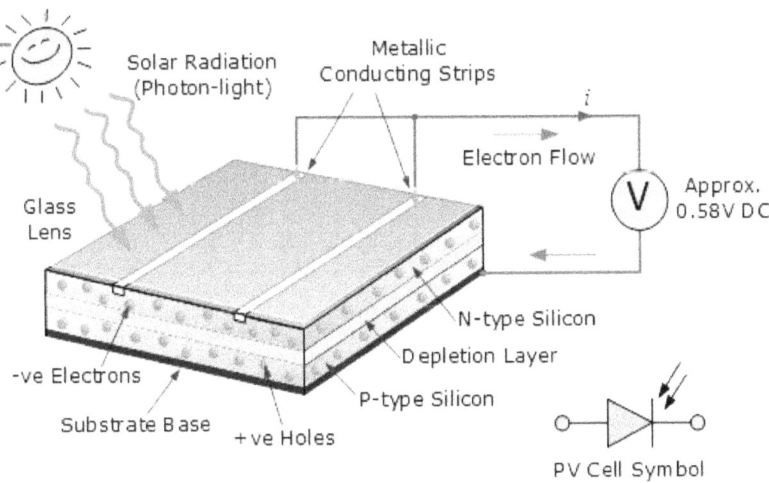

Solar Cell working - When Photons in Sunlight hit solar cell, electrons are freed and current starts flowing

When sunlight falls on a solar cell, the solar cell generates some voltage and current, which is called solar energy. A solar cell works like a torch battery in our house. Just as a battery, cell cannot power a whole house of lights and fans, so a single solar cell can not generate that much electricity. Therefore, a large solar panel is created by sticking many such solar cells on a special sheet and connecting them to each other with thin wires. This solar panel is called a solar panel. A solar array is created by connecting many solar panels to each other. Depending on your electricity needs, large amounts of solar energy are generated by connecting many solar panels or many solar arrays to each other. See Figure 1.

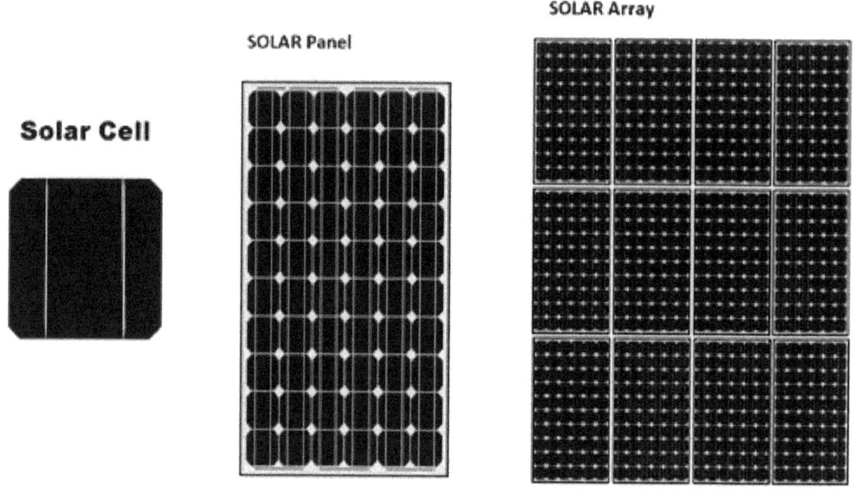

Front Side of the solar Panel

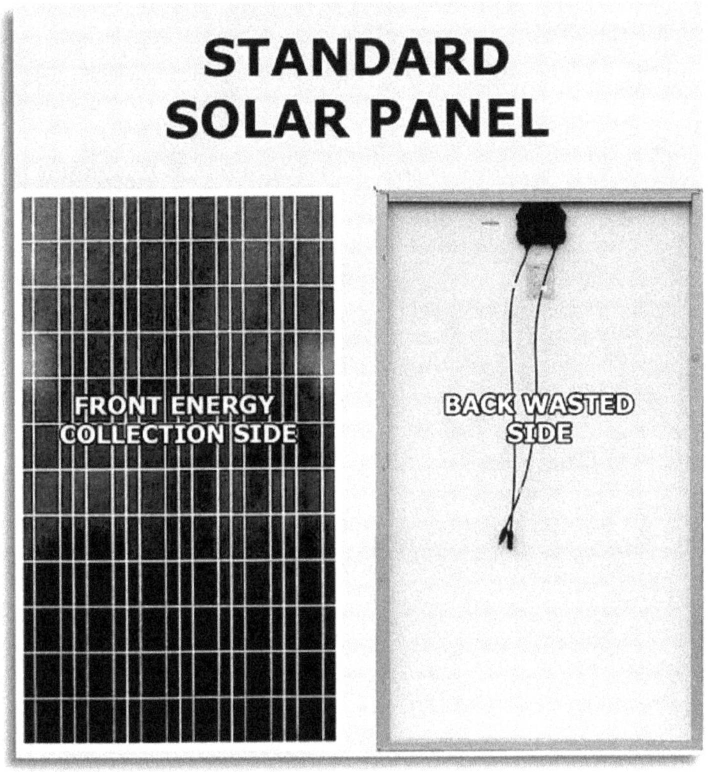

Backside of the Solar Panel

Large amounts of energy are generated by connecting many solar arrays together. Large amounts of electricity for an entire village or all the establishments in a large building can be generated by connecting solar panels together in this way. See the picture below.

Picture of a Solar Power Plant

CHAPTER THREE

BASIC ELECTRICITY

Let us learn a little about electricity so that we can understand how much electricity we need or how much electricity we need and how much solar energy we need to convert to meet that need. For this, we will have to learn some basic knowledge of Electrical Engineering. This information will be completely practical and we will have to use it further.

The electricity coming to our house comes from 3 wires. One is Phase, the second is Neutral and the third is Earthing. Let us call these three wires as P, N and E respectively. Out of these three wires, only two, P and N, come to our house from the electricity pole and the third wire is there to hold the P and N wires and this wire goes to the ground and is called E or Earthing. This E wire is used to prevent any appliance in our house from getting an electric shock. The current leaked from this wire goes to the ground. For this, a hole is made in the ground and a special rod is buried in the ground to connect the earthing wire i.e. wire E to it and the wire E is connected to that rod. See the figure. We will see how to do earthing in the next chapter.

Of these P and N, the electric current comes from the P wire and circulates in the equipment and returns through the N wire. See the figure.

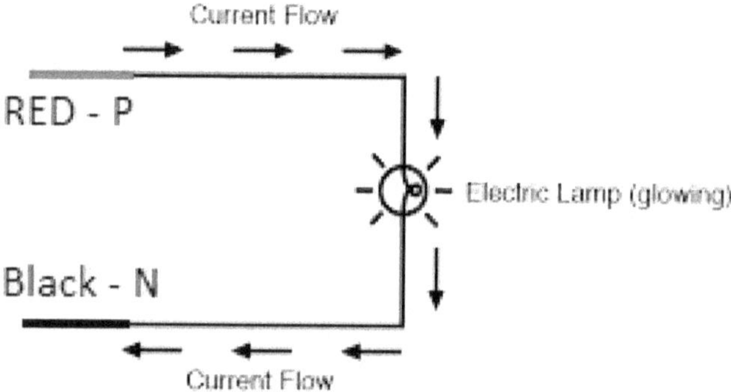

Flow of current

This electric current is measured in the unit of ampere (Amp). The electric current starting from P and returning from N is called current. Also, the electric charge between P and N is called voltage. The unit of voltage is volts. Both the words voltage and current are very important for us.

The electricity that comes to our house from the electricity pole is called alternating current or in short form AC. All the appliances in our house generally run on AC. The other type of electricity is called direct current or in short form DC. Now let's learn a little about DC. The flashlight battery in the house, power bank, inverter battery or car battery are all DC type appliances. In short, battery means DC type of electricity. Just like AC, DC also has two wires P and N, but they are called positive and negative. Positive is indicated as +ve and negative is indicated as -ve. This is indicated on many devices or on each battery as +ve and -ve. See the picture below

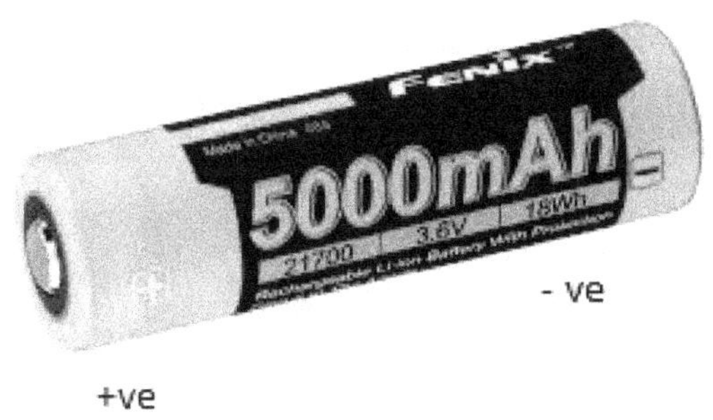

Pencil Cell is a source of DC

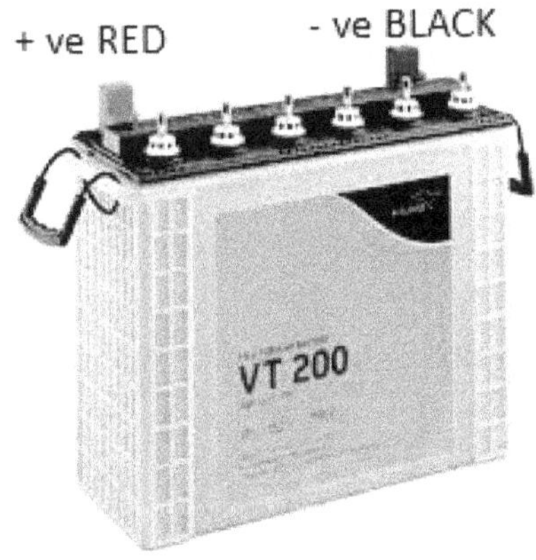

Lead Acid Battery

DC type of electricity can be stored. Every battery is a store of electricity just like we fill a bottle or jar with water, similarly electricity is stored in the battery and can be refilled when it is used up. That is what we call battery charging. When the car battery runs out many times, the car does not start, then the battery is charged and re-installed and the car starts.

The flow of AC type of electricity, i.e. AC current, is constantly changing, that is, its magnitude is constantly changing. Starting from 0, it reaches

+ve peak, decreases again to 0 and decreases further, reaches -ve peak, and increases again to 0 and again +ve peak, and this cycle continues until AC current starts. AC electricity cannot be stored and if it is not used at the moment it is generated, it is wasted. This is the biggest disadvantage of AC. AC current is represented by the symbol . See the picture.

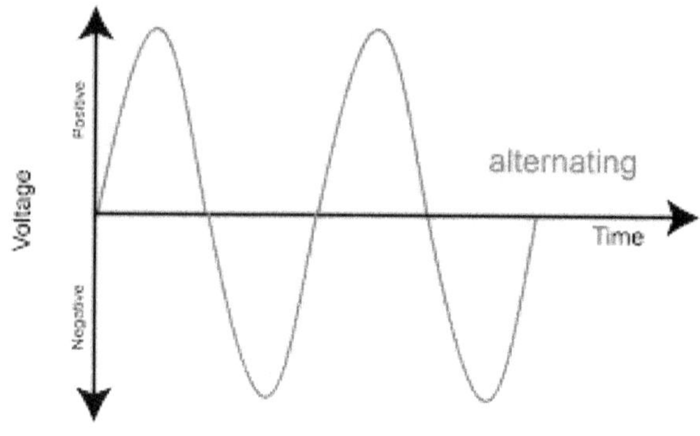

Symbol of AC

Also, DC current flows in the same amount or capacity. All sources of DC current have their capacity written on them, such as 12 volts on a battery or 3.7 volts if a mobile has a battery. Another unit used for batteries is called ampere hour (AH). We will learn more about it later. DC current is indicated by the symbol as shown below.

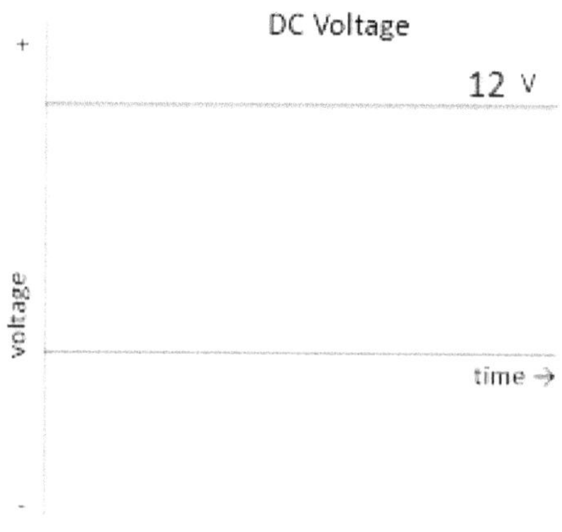

Symbol of DC

From now on, whenever we see the above symbols for AC or DC on any device, we understand whether the device is running on AC or DC.

Four things are important in terms of energy consumption, namely voltage, current, power and energy. Out of these, we have already identified voltage and current. Now let us identify power and energy.

Let us look at 4 important things in electrical engineering.

Voltage (V) Volts - Electrical load between 2 wires

Current (A) Amp - Electrical current in the wire

Power (W) watts=V x I - Electrical power or electrical power required to run the device

Energy (E) W x hours - Power x time. How much time we used the power

For example - Let us take the example of a steam iron. The iron is usually of 500 watts. That is, the power (watts) of the iron is 500 watts. The voltage of the electricity coming to our house is AC 230 v, that is, the iron is an appliance that runs on AC 230 v. Now let us see how much current and energy the iron consumes.

W = V x I means 500 w = 230 v X I amp

That is, I = 500 / 230 = 2.17 Amp

This means that the power of the iron is 500 watts and it consumes a current of 2.17 A at AC 230 volts. Suppose we run this iron for 4 hours, let's see how much energy is consumed.

Energy E = watt x hours i.e. 500 watt X 4 hours = 2000 WH

The electricity bill that your electricity supplier (MSEB or now its name is MSEDCL) sends you, shows how many units of energy you have consumed per month and accordingly you get a bill. The meaning of this unit is given below -

1 unit = 1000 watt X 1 hour i.e. if you run 1 appliance of 1000 watt power for 1 hour, 1 unit of energy is consumed and accordingly you get a bill.

Now let's see how much energy we used in the above example of the iron and how much the bill can be.

Energy E = watt x hours i.e. 500 watt X 4 hours = 2000 WH

That is, E = 2000 WH = 2000/1000 = 2 KWH = 2 units

This means that if we run an iron of 500 watts power running on AC 230 v for 4 hours, the iron will consume 2.17 A current and 2 units of energy will be consumed.

Summary - Let's look at another example to understand voltage, current, power and energy properly. The width of the pipe of a water pump is the current, while the speed of the water thrown out is the voltage. The time taken to fill the tank depends on both the width of the pipe (current) and the speed of the water coming out (voltage), so its power i.e. watts (W = V x I) is written on any appliance. Watt is a unit of power and the product of how much time we use this power (E = watt x hours) is called energy. If 1000 watts of power is used for 1 hour, 1 unit of energy is consumed.

Sometimes watts are not written on the appliance and current and voltage are written, then we can calculate the watts required by that appliance.

For example, It is written as V = 230 V, I = 2 Amp on any appliance then, we can say that this appliance has a power of w = V X I = 230 X 2 = 460 watts.

In the following case, we will use the terms V, I, W and E with their letters for convenience.

To each electricity consumer, the electricity distribution company sends a bill according to his/her energy wattage. Generally, the electricity bills of all the electricity distribution companies, MSEDCL, BEST, TATA POWER or ADANI POWER in Mumbai are prepared in more or less the same way. If you study your bill a little, you will notice how you get the bill. Let us study the bill of MSEDCL.

All the details in this bill are generally understandable, but if you look at the back of the bill or page 2 of the bill above , you will notice that many technical things are shown there. Look at the photo of the bill given below. Look at the second / back page of it.

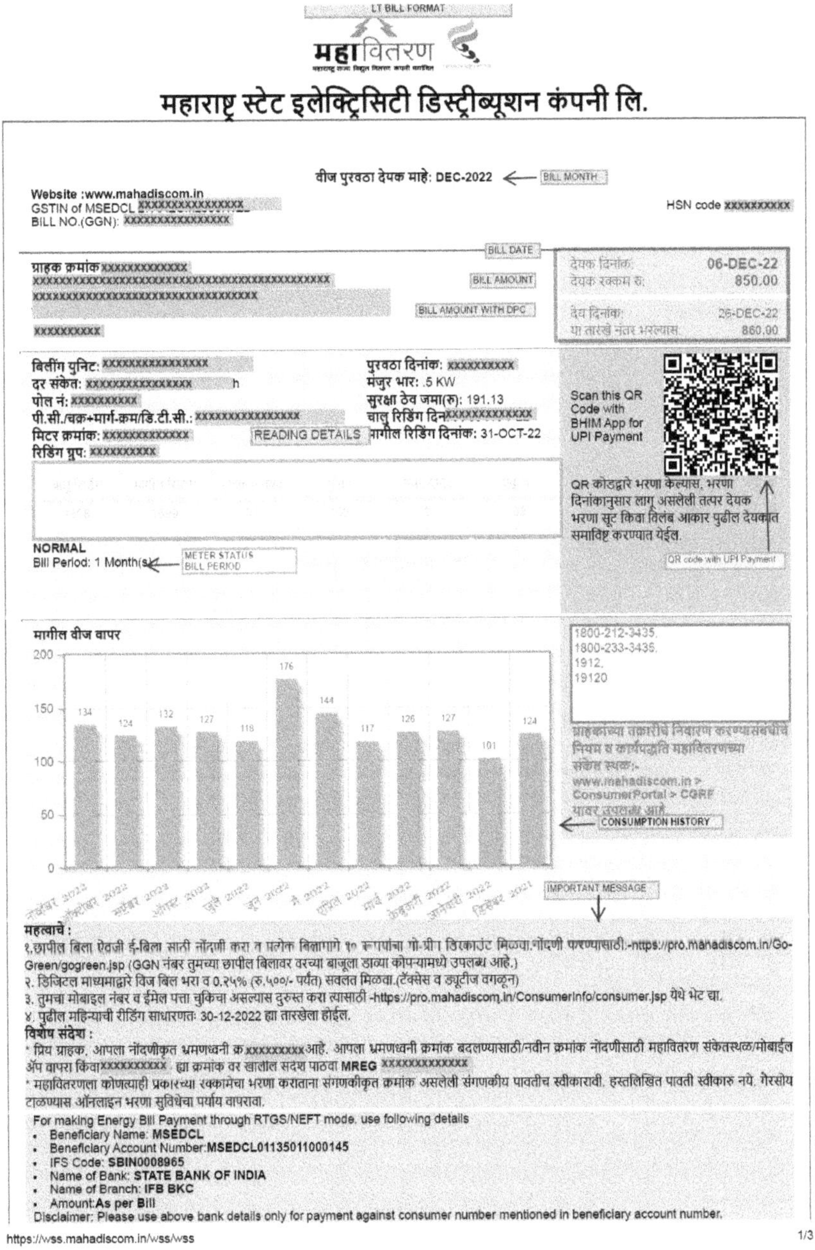

LT BILL FORMAT

महावितरण

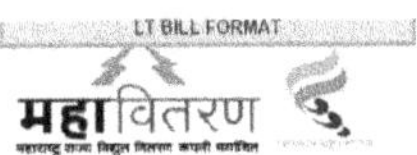

महाराष्ट्र स्टेट इलेक्ट्रिसिटी डिस्ट्रीब्यूशन कंपनी लि.

वीज पुरवठा देयक माहे: DEC-2022 ← BILL MONTH

Website :www.mahadiscom.in
GSTIN of MSEDCL xxxxxxxxxxxxxxx
BILL NO.(GGN): xxxxxxxxxxxxxxxx

HSN code xxxxxxxxxx

BILL DATE

ग्राहक क्रमांक xxxxxxxxxxxx
xxxxxxxxxxxxxxxxxxxxxxxxxxxxxxxxxxxxxxxxxxxx
xxxxxxxxxxxxxxxxxxxxxxxxxxxxxxxxxx
xxxxxxxxxx

BILL AMOUNT

BILL AMOUNT WITH DPC

देयक दिनांक:	06-DEC-22
देयक रक्कम रु:	850.00
देय दिनांक:	26-DEC-22
या तारखे नंतर भरल्यास	860.00

बिलींग युनिट: xxxxxxxxxxxxxxxx
दर संकेत: xxxxxxxxxxxxxxxx h
पोल नं: xxxxxxxxxx
पी.सी./चक्र+मार्ग-क्रम/डि.टी.सी.: xxxxxxxxxxxxxxxx
मिटर क्रमांक: xxxxxxxxxxxxx
रिडिंग ग्रुप: xxxxxxxxxx

पुरवठा दिनांक: xxxxxxxxxx
मंजुर भार: .5 KW
सुरक्षा ठेव जमा(रु): 191.13
चालु रिडिंग दिनxxxxxxxxxxxx
मागील रिडिंग दिनांक: 31-OCT-22

READING DETAILS

NORMAL
Bill Period: 1 Month(s) ← METER STATUS BILL PERIOD

Scan this QR Code with BHIM App for UPI Payment

QR कोडद्वारे भरणा केल्यास, भरणा दिनांकानुसार लागू असलेली तत्पर देयक भरणा सूट किवा विलंब आकार पुढील देयकात समाविष्ट करण्यात येईल.

QR code with UPI Payment

मागील वीज वापर

नोव्हेंबर 2022	ऑक्टोबर 2022	सप्टेंबर 2022	ऑगस्ट 2022	जुलै 2022	जून 2022	मे 2022	एप्रिल 2022	मार्च 2022	फेब्रुवारी 2022	जानेवारी 2022	डिसेंबर 2021
134	124	132	127	118	176	144	117	126	127	101	124

1800-212-3435,
1800-233-3435,
1912,
19120

ग्राहकांच्या तक्रारींचे निवारण करण्यासंबंधीचे नियम व कार्यपद्धति महावितरणच्या संकेत स्थळ:- www.mahadiscom.in > ConsumerPortal > CGRF यावर उपलब्ध आहे.

← CONSUMPTION HISTORY

IMPORTANT MESSAGE

महत्वाचे :
१.छापील बिला ऐवजी ई-बिला साठी नोंदणी करा व प्रत्येक बिलामागे १० रूपयांचा गो-ग्रीन डिस्काउंट मिळवा.नोंदणी करण्यासाठी.-https://pro.mahadiscom.in/Go-Green/gogreen.jsp (GGN नंबर तुमच्या छापील बिलावर वरच्या बाजूला डाव्या कोपऱ्यामध्ये उपलब्ध आहे.)
२. डिजिटल माध्यमाद्वारे विज बिल भरा व 0.२५% (रु.५००/- पर्यंत) सवलत मिळवा.(टॅक्सेस व ड्यूटीज वगळून)
३. तुमचा मोबाइल नंबर व ईमेल पत्ता चुकिचा असल्यास दुरुस्त करा त्यासाठी -https://pro.mahadiscom.in/ConsumerInfo/consumer.jsp येथे भेट द्या.
४. पुढील महिन्याची रीडिंग साधारणतः 30-12-2022 ह्या तारखेला होईल.

विशेष संदेश :
* प्रिय ग्राहक, आपला नोंदणीकृत भ्रमणध्वनी क्र.xxxxxxxxxआहे. आपला भ्रमणध्वनी क्रमांक बदलण्यासाठी/नवीन क्रमांक नोंदणीसाठी महावितरण संकेतस्थळ/मोबाईल ॲप वापरा किंवाxxxxxxxxxx . ह्या क्रमांक वर खालील संदेश पाठवा **MREG** xxxxxxxxxxxxx
* महावितरणला कोणत्याही प्रकारच्या रक्कमेचा भरणा करताना संगणकीकृत क्रमांक असलेली संगणकीय पावतीच स्वीकारावी. हस्तलिखित पावती स्वीकारु नये. गैरसोय टाळण्यास ऑनलाइन भरणा सुविधेचा पर्याय वापरावा.

For making Energy Bill Payment through RTGS/NEFT mode, use following details
- Beneficiary Name: **MSEDCL**
- Beneficiary Account Number:**MSEDCL01135011000145**
- IFS Code: **SBIN0008965**
- Name of Bank: **STATE BANK OF INDIA**
- Name of Branch: **IFB BKC**
- Amount:**As per Bill**

Disclaimer: Please use above bank details only for payment against consumer number mentioned in beneficiary account number.

https://wss.mahadiscom.in/wss/wss 1/3

Front page of MSEDCL monthly Bill

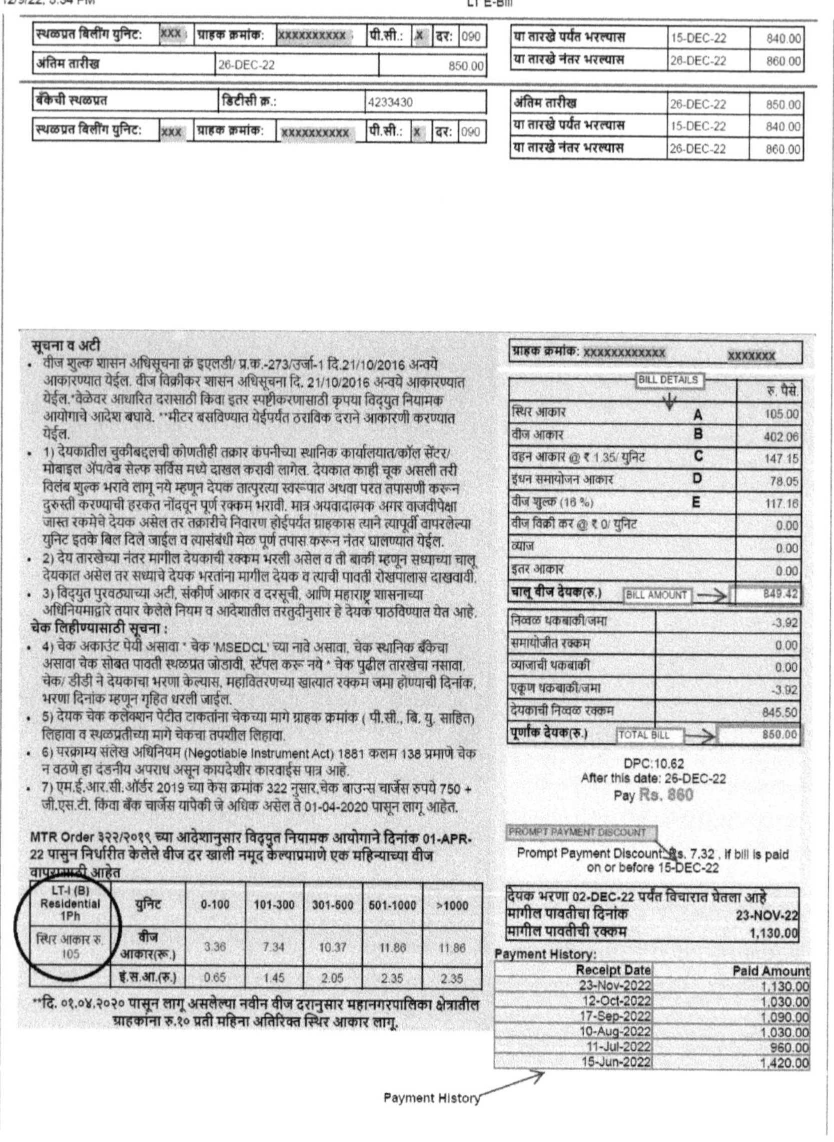

Back Page of MSEDCL monthly Bill

A - This fixed amount is determined by the type of your connection like LT I i.e. for home or residential connection Rs.105/-

B- Electricity amount is calculated as follows. Total 109 units

= 100 units x 3.36 rate + 9 units X 7.34 rate

= 336+66.06 = 402.06

C - 109 X 1.35 = 147.15 Electricity comes to your house from the National Grid. This charge is levied for the cost of transporting it from other places.

D - Fuel adjustment amount is calculated as follows. Total 109 units

= 100 units x 0.65 rate + 9 units X 1.45 rate

= 65 + 13.05 = 78.05

E - Electricity charge is 16%.

(A + B + C + D) 16% = 117.16

The fuel adjustment charge is very interesting. We mainly get electricity from thermal power stations, that is, by burning coal! Since good quality coal is not available in the top layer of the earth these days, it has to be dug deeper, so consumers have to pay this fuel adjustment amount under the name of fuel adjustment.

CHAPTER FOUR

ELECTRICAL SAFETY

Electric shock place

1) If a person stands on the ground and any part of the body comes in contact with a live line, that person gets an electric shock.

2) Even if insulated from the ground, if two wires of different polarity of the power supply line or two lines of different voltages come into direct contact, an electric shock occurs.

3) If the earthing connected to the metal body of an electric machine is not proper and the insulation resistance of such a body is weak, an electric shock occurs if the body comes into contact with that body.

4) If someone else touches the person in contact with electricity without taking adequate protection, the person gets electrocuted.

In this way, due to electric shock, the person is pushed away with force and becomes unconscious, or sometimes he can get electrocuted and die. Therefore, the wireman/electrician should work very carefully while working on electricity.

Precautions to be taken to avoid electric shock:

1) Always use insulated tools while working on electricity.

2) If the wire of an appliance is damaged, do not work without the wire.

3) If the covers on the switch, plug holder, etc. are cracked, they should be replaced immediately.

4) Earth the metal body of the appliance.

5) Do not work on a live line.

6) Do not pour water to extinguish a fire caused by electricity.

7) Do not use your hand to check for the presence of electricity.

8) After switching off the switch of an installation or point, do not work there without being completely sure that the electricity supply has been switched off.

9) Wrap insulation tape on the wire joints.

10) Turn off the main switch while changing the fuse wire.

Precautions to be taken while removing a person who has come into contact with electricity

If someone comes into contact with electricity, immediately turn off the power supply to that place. If this is not possible, remove the person from the live electrical part by wearing rubber gloves. If rubber gloves are not available, hold them with a dry cloth and push them away. Dry wood can also be used for this purpose, but under no circumstances should you pull the person directly by the hand, as you may get a shock if you are not insulated.

Precautions to be taken after removing the casualty from the electric shock.

After removing the casualty from the electric shock, first aid should be given before the doctor arrives and the following precautions should be taken -

1) Keep the casualty in an open area

2) Loosen the clothes if necessary

3) If there is an injury, keep the injured part in a comfortable position and cover the wound

4) If the casualty is unconscious, use artificial respiration

5) Give him a drink like tea or coffee if possible after regaining consciousness

6) The person treating the casualty should not talk to him as if he has made a mistake

7) If any legal action is required, inform the authorities

Precautions to be taken while working in the electrical field

1) Before starting any work, turn off the main switch of the circuit and remove the fuse in it in a safe place

2) Always connect the live wire (+ve) of the circuit through the switch

3) While working on a ladder, the ladder should be tilted slightly and not kept straight Take the help of a helper to hold it

4) Do not supply electricity or turn on the equipment without having complete knowledge of the equipment

5) Do not work standing in a damp place

6) Before repairing or opening any equipment, remove its diagram

7) Do not move the electrical equipment from one place to another while it is running

8) Fuses should be installed of the right size

9) Do not work with excessive confidence, that is, work with all caution
10) Do not work if you are mentally stressed
11) Do not bring a burning flame into the battery charging room.

CHAPTER FIVE

TYPES OF SOLAR SYSTEMS

In the previous article, we learned how the use of solar energy is essential and beneficial. In this article, we will see the types of solar energy systems. We will be able to decide which type of system is right for us. In the next article, we will see how we can design this system according to the electrical load in our home, office or factory.

We use solar energy to meet our electricity needs. There are 3 types of systems to generate solar energy. We can make that system by deciding which one is right for us. In fact, it can be very cheap if we design and make our own system and that is why this article is for us.

3 Types of Solar Energy Systems -

1) On Grid System 2) Off Grid System 3) Hybrid System

Let us understand these 3 systems one by one. Where electricity is not continuous or goes out frequently, hybrid system is more beneficial. Therefore, hybrid systems are more commonly used in rural areas. Also, on-grid systems are more suitable for large capacity plants. We will see that in the next section. A diagram is given in the information of each system from which you can easily understand that system.

1) ON-GRID system - Grid means network. All the power generating sets in our country are connected to each other by electricity wires and a network of them has been created. The electricity that we get in our home or office from the electricity supply company comes from this grid. If there is a shortage of electricity in a state or if there is a high demand, electricity is supplied from other generating sets or from generating sets in other states to the state where there is a shortage through this network of electricity wires. This is the biggest advantage of this network or grid. If the electricity generated in the power generating set is not used, it goes to waste because it cannot be stored. So now from the diagram below, you will understand how the on-grid system should generally work.

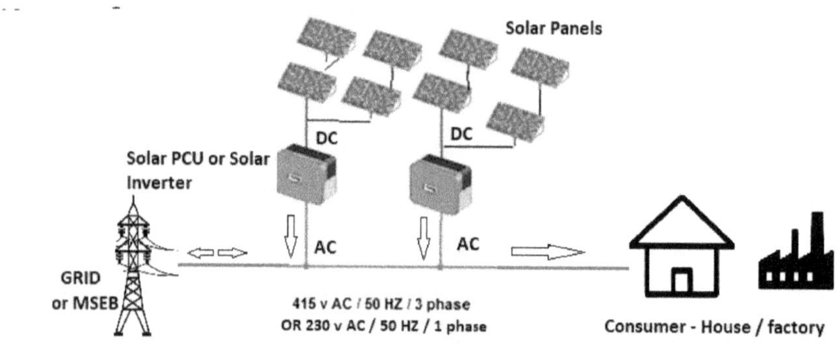

Schematic Diagram of ON GRID system

The energy generated from the panels is of DC type. For our domestic use, it should be of AC type because all the appliances in our house or office, for example, lights, fans, TV and fridges, etc., run on AC. The work of converting this DC into AC is done by the Solar PCU (Power Conditioning Unit) or in simple terms, the Solar Inverter. The electrical connection of this system is as shown in the figure. As shown in the figure below, a house or factory is running on solar energy. If the electricity requirement of this house or factory is more than the solar energy generated, the remaining electricity requirement is met by taking electricity from the power supply company, i.e. from the grid. If the solar energy generated is more than the requirement, the excess energy is returned to the power supply company or the grid and the customer gets paid as per its unit rate. This is called an on-grid system. To measure the energy received and supplied by the power company, your existing electric meter is replaced and a new net meter or bi-directional meter is installed. This meter shows you the net (net) number of units of electricity you have used, so it is called a net meter. It is not necessary to install a net meter to install an on-grid system. But since the excess electricity generated can or does go for free, both on-grid systems and net meters are installed. The decision whether to install a net meter or not is made according to the capacity of your solar system.

Let us look at the situation of solar energy generation and load at different scenarios in the following 4 figures.

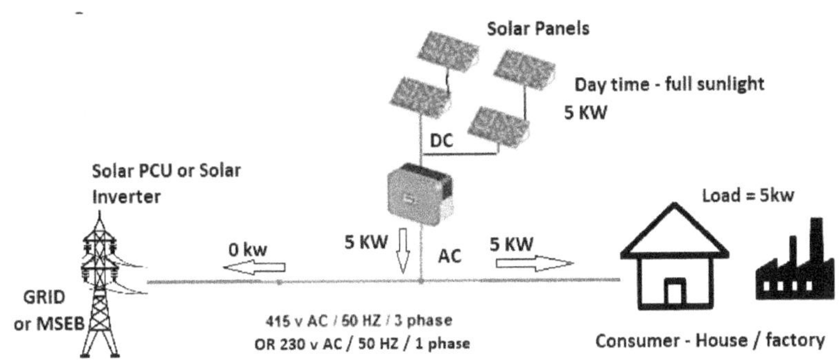

Condition 1 - Generation is same as LOAD

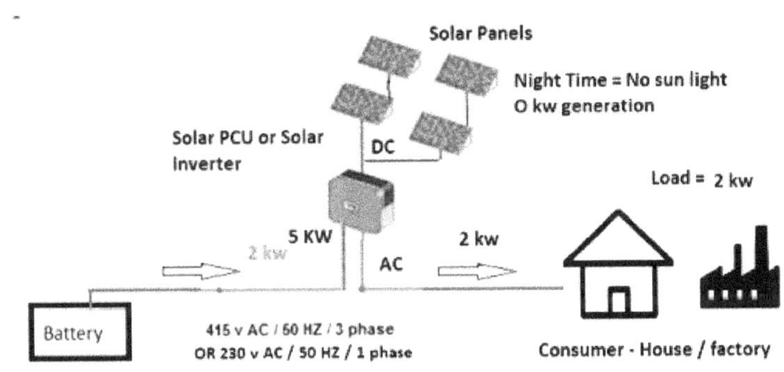

Condition 2 - Generation is more than LOAD

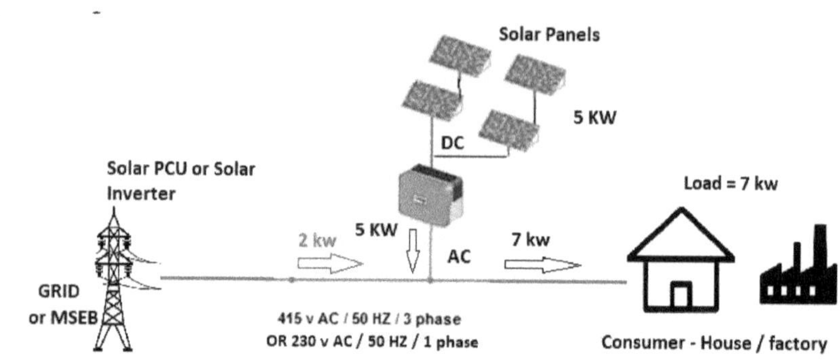

Condition 3 - Generation is less than LOAD

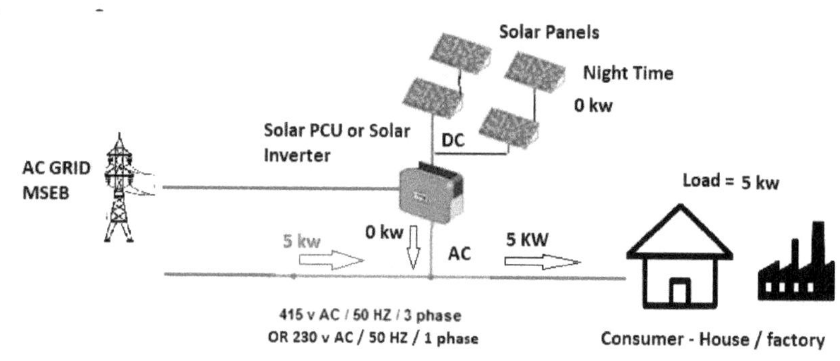

Generation is zero (night times) & LOAD is full

Advantages and disadvantages of on-grid systems -

Advantages -

1) Generate large amounts of solar energy (5 kw or more) and earn money directly by returning it to the power company or grid.

2) Generate and use solar electricity for your needs and greatly reduce your electricity bill from the power company.

3) Every 1 kw of solar electricity generated saves 900 grams of carbon emissions per hour and helps the environment.

4) You can reduce your dependence on the power company.

Disadvantages -

1) It is economical to generate on a large scale.

2) Since electricity cannot be stored in an on-grid system, it is only useful for the day.

3) You have to go through a complicated process to install a net meter with the power company.

2) OFF-GRID System - Off Grid System means where grid power or electricity from the electricity company is not available and the solar system used there is called off grid system. Many times we experience that if we want electricity connection in our garden or field and other remote areas, MSEB tells us that we will have to put so many poles and it will cost so many lakhs. This is because electricity does not reach that place. At such times we can generate solar energy. We can store the electricity generated more than our requirement during the day in the battery and use it after sunset or at night. Such a system is called off grid system. These types of systems have become very useful in rural areas or places where electricity has not reached. These systems are a little expensive but if the electricity generated is used carefully, it becomes very useful and economical. From the figure given below, you can see how off grid system works. In the next article, we will see how to design this system to be useful to you.

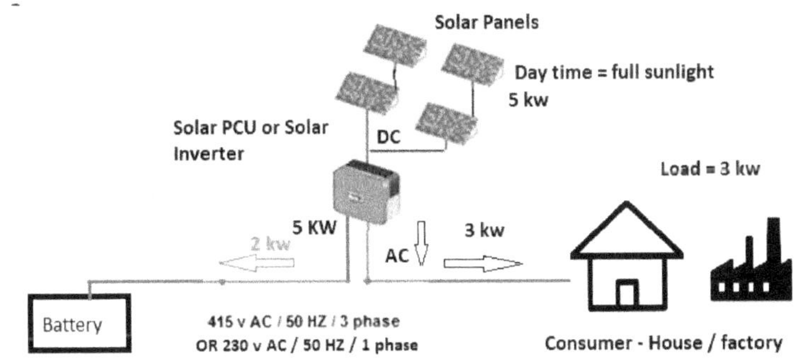

Generation more than LOAD is stored in BATTERY

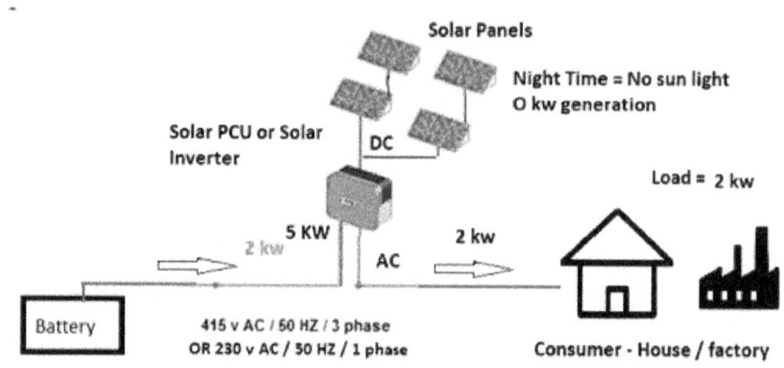

Generation is less than LOAD and diffrential current is drawn from BATTERY

HYBRID System -

This system is a combination of both on-grid and off-grid systems. As we have seen, on-grid is used where electricity is always available, off-grid is used where electricity is not available, and hybrid systems are used where electricity is available but is interrupted many times. Hybrid systems are also more useful because in our country, except for major cities, electricity is interrupted frequently in many places. There are more options available in hybrid systems, such as setting the system according to your needs or the availability of electricity. First, we can set the sequence of solar energy - battery - grid AC or solar energy - grid AC - battery or any combination of these three power sources. We can decide this setting according to our needs, the frequency of power interruptions and its duration. Due to this feature of hybrid systems, hybrid systems can be a viable and economical alternative to electricity supplied by the electricity supply company in semi-urban and rural areas.

We can understand the working of a hybrid system from the diagrams below.

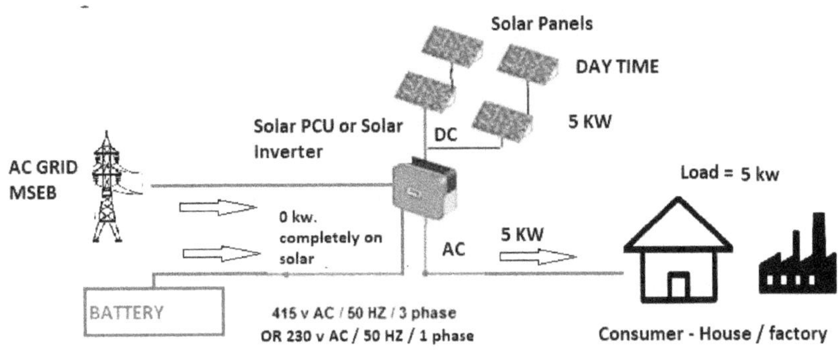

Generation is same as LOAD so no current is drawn from GRID & BATTERY

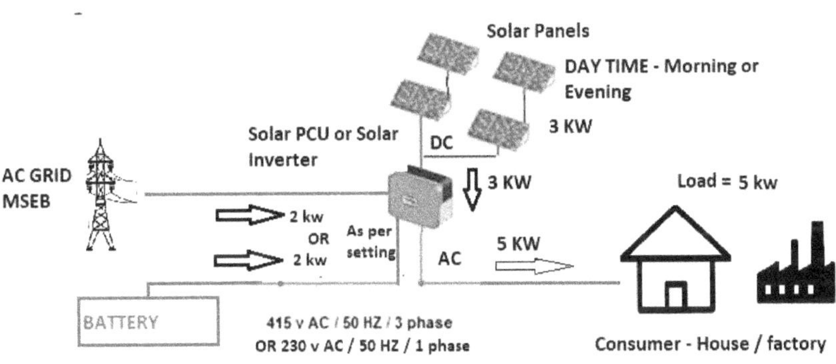

Generation is less than LOAD so differential current is drawn from either GRID or BATTERY dependingon the setting set on INVERTER

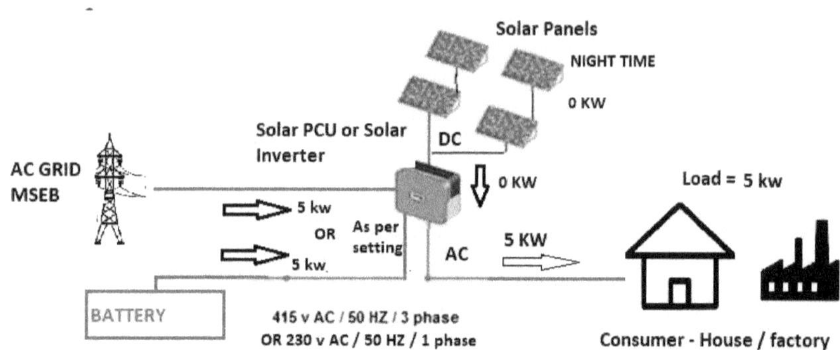

Generation is ZERO with LOAD on. So differential current is drawn from either GRID or BATTERY dependingon the setting set on INVERTER

COMPONENTS OF THE SOLAR SYSTEMS

COMPONENTS OF SOLAR SYSTEMS

Solar system is made up of 5 parts. On grid system is made up of four parts because it does not have a battery. We will now see those parts and their information.

1) Solar panels
2) Solar inverter
3) Battery
4) Structure for installing solar panels
5) Electrical wires, cables and connectors

Now we will get to know about all the components one by one.

1. **Solar panels –**

Solar panels or also called PV panels i.e. Photo Voltaic Panels. However, we will use the simple term solar panels. Solar panels are made by connecting many solar cells to each other. Each solar cell generates some voltage. A solar panel is made by connecting these solar cells in series or parallel connection. Each solar cell generally generates 0.5 volts. See the diagram of solar cell and solar panel.

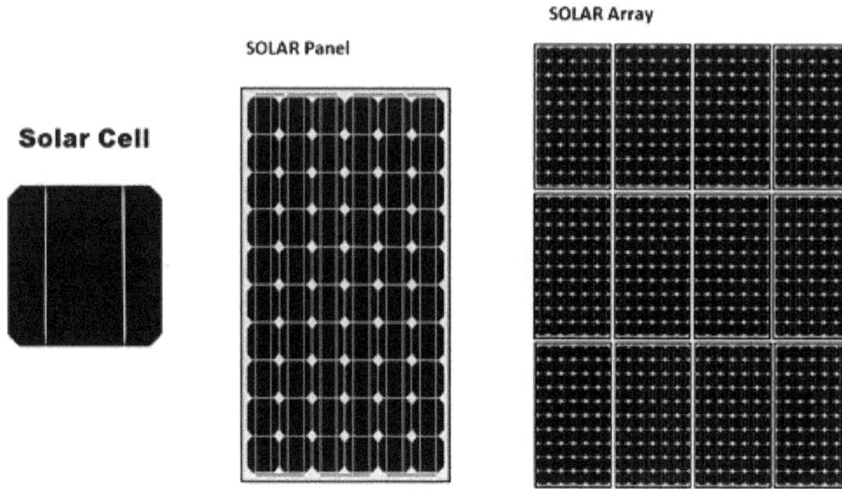

Enter Caption

A solar panel is made by connecting 30, 36, 60 or 72 solar cells in series. A 30-cell panel produces 15 volts, while a 60-cell panel produces 30 volts. Similarly, a 36-cell panel produces 18 volts and a 72-cell panel produces 36 volts. A panel made by connecting 30 or 36 cells is called a 12-volt panel and a panel made by connecting 60 or 72 cells is called a 24-volt panel. The reason for this is that 12-volt panels are used to charge a 12-volt battery and 24-volt panels are used to charge a 24-volt battery or two 12-volt batteries. These solar panels also do the work of charging batteries in off-grid and hybrid systems, so solar panels are made in 12 volts or 24 volts. There is another dimension to solar panels and that is how much power the solar panel has. The unit of power is Watt. Solar panels are available in various capacities from 6 watts to 370 watts. Panels of 6 watts to 15 watts generate 6 volts. We will consider 12 volts and 24 volts panels here. Solar panels from 20 watts to 150 watts are of 12 volts and solar panels from 150 watts to 370 watts are of 24 volts. While designing a solar system, we have to decide which solar panels we are using. We will see that in the following articles. When we say a 370 watt solar panel, it generates a maximum of 370 watts, that is, it generates 370 watts in the afternoon and its watts are less in the morning and evening. Therefore, a solar panel is expressed in 2 units. For example, a solar panel is described as 370 Wp (370 Watts Peak), 24 volts

(24 volts). Panels of different capacities are used for different applications, such as 10Wp-6V for solar lanterns, 36Wp-12V for solar street lights, and panels of 200Wp-24V or more are used for solar systems.

In a solar system, a system of as much power as we need is designed by connecting many solar panels in series and parallel. For example, 8 panels of 250 Wp are connected for a 2 kw system or 12 panels for 3 KW. In summary, it can be said that we can generate the power we want by connecting solar panels to each other. We will see more technical details about this in the next chapter on system design.

2. Solar Inverter –

This device is sometimes also known as a power conditioning unit. Let us call it a solar inverter. As we saw in the previous chapter, just as there are three types of solar or solar systems, there are also three types of solar inverters. In fact, the types of solar systems are determined based on this type of solar inverter. Solar inverters are also available in 3 types.

2.1) On Grid

2.2) Off Grid

2.3) Hybrid

Let us look at the working of all these inverters from the point of view of their connection points. While looking at the diagram of each inverter, keep the inverter in the middle and look at the left side as input and the right side as output, so that it will be easy to understand.

2.1) **ON GRID INVERTER** - This type of inverter has 6 connection points. 2 connection points are for connecting the +ve and -ve wires coming from the solar panel, 2 are for connecting the 2 wires coming from the electricity pole i.e. the main connection, the input, AC phase and AC Neutral, while the remaining 2 points are for connecting the output AC phase and AC Neutral. The first 4 connection points are for input and the remaining 2 are for output. There is no provision for connecting a battery in an on-grid inverter. If you generate more electricity than the load you have or more than your requirement, you can give this excess electricity back to the power company, so this system is called an on-grid system. This inverter is directly connected to the grid, so it is called an on-grid system. In this way, to give electricity back to the grid or to the power company, you have to apply to the power company and once it is approved, your meter has to be changed. The new meter is of the net meter type which measures the energy flowing in both directions. The net meter shows both the readings of the electricity you have taken from the electricity company and the electricity you have given to the electricity company, and the customer gets the electricity bill by subtracting them.

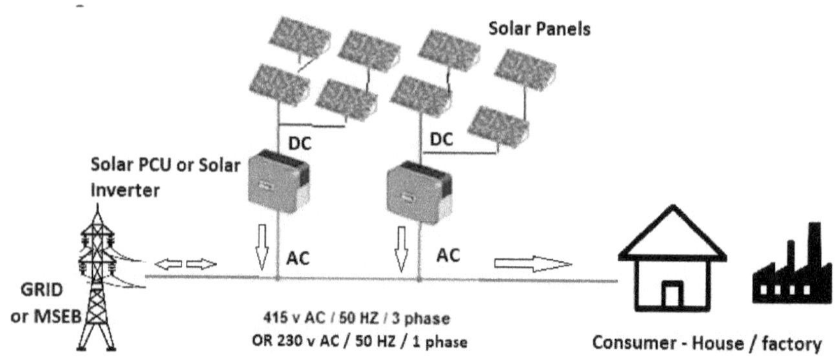

Schematic Diagram of ON GRID Inverter

2.2) OFF GRID INVERTER

The difference between on grid and off grid inverter is in its name. Off grid inverters are used where there is no electricity from the power company or grid. This type of inverter also has 6 connection points. 2 points are for solar panels, 2 points are for batteries instead of grid or AC input and the remaining 2 points are for AC output. In this type of system, more electricity is generated than the load. During the day, the load runs completely on solar energy and the excess energy generated is stored in the battery. After sunset, when solar energy generation stops, the battery supplies energy to the load and the load runs.

The capacity of the battery is determined by how much load is there after sunset and how long it is to run, and accordingly, a battery of that capacity is connected to the off grid inverter. Battery capacity is expressed in AH. The number of solar panels is determined on this AH unit or capacity of the battery. The number of solar panels and batteries is designed in such a way that during the day the load runs completely on the solar panels and the batteries are also fully charged. All the load runs on the batteries in the evening. Thus, off-grid inverters are used where electricity is not available at all. Since the battery is an essential part of this system, this type of system becomes expensive. We will see more details about this system in the next chapter.

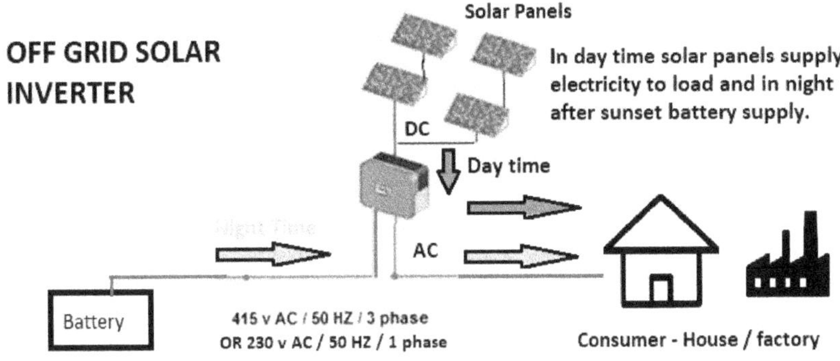

Schematic Diagram of OFF GRID Inverter

2.3) HYBRID Inverter -

A hybrid system is basically an off-grid system, but hybrid is a combination of on-grid and off-grid systems. A hybrid system is used in places where there is electricity but the power supply is not consistent or the power goes in and out frequently. This system is different from on-grid or off-grid in the number of connection points. This system has 8 connection points. 2 for AC input, 2 for solar panels, 2 for batteries and 2 for AC output.

In this system, less solar panels can be used than in an off-grid system because when AC is available, AC can also be used to charge the batteries. In this system, we can set the inverter according to the times and duration of our load shedding and according to our battery capacity and achieve the best possible output. For example - if load shedding occurs every afternoon, we can set the battery to charge on solar panels. Similarly, if load shedding occurs in the evening, we can set the battery to charge from AC.

Also, in this type of inverter, we can set the output as well. If load shedding is for a long time, we can set the output to use solar power first, then AC if solar power is not available, and then use battery if both solar and AC are not available. In all these respects, hybrid inverters are very useful and powerful options in areas where power goes out frequently.

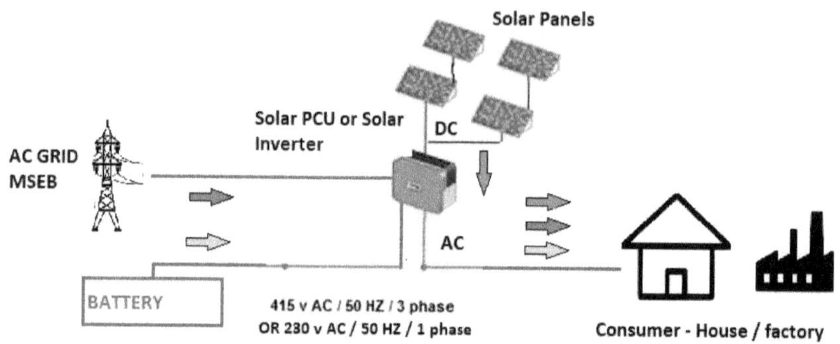

Schematic Diagram of HYBRID Inverter

3) Battery –

Battery is not a new thing in our country, we run many things from household torches to home inverters on batteries. The main function of a battery is to store energy. As we have seen in the previous case, AC electricity cannot be stored but DC electricity can be stored. Since the electricity generated by solar panels is of DC type, we can store it, that is, we can store solar or solar energy in batteries. The main function of a battery is to store energy. A battery is described in terms of its voltage and current. For example, any battery is described as 12 volts, 7 AH or 12 volts, 26 AH or 100 AH, 150 AH, 180 AH or 200 AH. Now let us see what AH is. AH stands for Ampere Hour. Current is measured in amperes and is represented by the letter A and time is measured in hours and is represented by the letter H. Thus, the specification of the battery is completed with both the units V and AH. Let us understand it with the following example.

12 V, 100 AH battery - This means, this battery can provide 100 Amp current at 12 v for 1 hour or

50 Amp current at 12 v for 2 hours or

10 Amp current at 12 v for 10 hours or

1 Amp current at 12 v for 100 hours.

AH simply means that the product of the current the battery provides and the number of times it allows it to do so is Amp x Hour = AH. That is, AH is the capacity of the battery. In reality, when using the battery, this calculation changes a little, that is, the battery capacity time obtained by the calculation is slightly reduced. We will learn more about it in the next chapter.

There are 2 types of batteries. One is called Lead Acid and the other is called SMF -Sealed Maintenance Free. There are new types in the SMF type such as Lithium Ion (Li Ion), Lithium Phosphate. (Li FeSo4), both of these battery changes we will see in more detail in the next chapter.

Lead acid type batteries need to be filled with pure water at regular intervals, which is called battery water or distilled water. The battery contains a solution of sulfuric acid, in which this water is mixed. Lead acid

Batteries are high capacity and can supply current for a long time. SMF type batteries are generally of low capacity and size but are very expensive. The energy-giving capacity of any battery usually decreases after 5 years and therefore the battery has to be replaced after 5 years. Therefore, off-grid or hybrid systems using batteries are expensive because it includes the cost of the battery and they cost money even after 5 years, but where electricity goes out frequently, off-grid or hybrid systems are very useful, in fact, an electricity-efficient solution.

4) Structure -

In a solar system, a system of as much power as we need is designed by connecting many solar panels in series and parallel. For example, 8 panels of 250 Wp are connected for a 2 kw system or 12 panels for 3 KW. In summary, it can be said that we generate the power we need by connecting the solar panels to each other.

All these solar panels have to be connected to each other by placing them in a specific direction and angle. We will see later how many degrees of angle and in which direction they have to be placed. See the figure below and you will see that a strong iron structure is required to erect solar panels in this way. This iron structure is made of galvanized iron because galvanizing is done on the iron to prevent it from rusting in the rain. See the pictures below, different types of structures or structures have to be used.

See pictures of various structures on the next page.

Ground Mounted Structure

Structure normally used for Solar Pumps

The solar panel generates the most energy when the sunlight falls on the solar panel at 90 degrees, so you have to erect the solar panel at some angle. The structure also has to be such that the solar panel can be erected at such an angle. See the figure below.

Due to the following reasons, the sun rays that emanate from the sun and are perpendicular (at 90 degrees) to the solar panel generate the most energy. The rays that do not fall on the solar panel also generate energy but to a lesser extent.

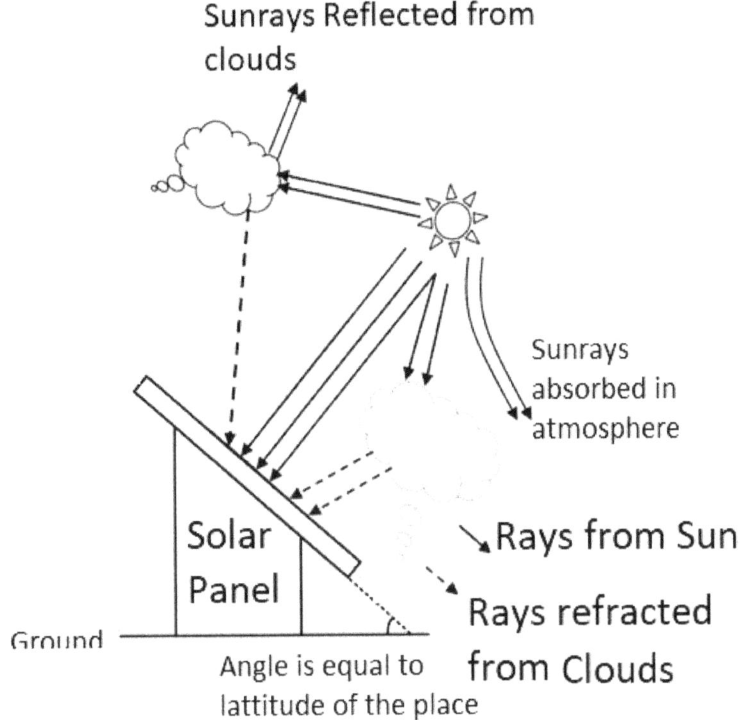

Importance of Angle of the installed panel

In our everyday language, the sun moves from east to west on the earth or technically the earth rotates on its axis from west to east, so we see the sun rising in the east and setting in the west every day. Since the earth's axis is slightly tilted, we have to take the angle of the solar panels as per the longitude at which we are installing them.

For example, if we are installing solar panels in Mumbai, the longitude of Mumbai is 18.970 north and the latitude is 72.820 east, so the angle of the solar panels will be 18.970 from the ground. Actually, this angle is a little less and more and changes according to the season, but to make things easier, we can install solar panels by taking this angle as the longitude. Also, this information is for places in the northern hemisphere like India. It should be noted that the math will be different for places in the southern hemisphere.

As shown in the figure, the iron structure that needs to be made to install solar panels is called a structure. This type of structure needs to be made keeping in mind both the number of solar panels and the free space. Such a structure or structure is also made in iron or aluminum. If it is an iron structure, red oxide and silver metallic color are applied on it. Due to red oxide and silver metallic color, the structure or this structure remains safe from corrosion. If GI (Galvanized Ion) is used, the structure remains safe from corrosion for a longer time, but GI is expensive.

Many photos of the structure are given after the reference list. All these photos can be used as a guide. The structure can be made in many ways, at your convenience, just follow the principle that the angle from the ground should be correct.

5) Wiring and Connectors –

The wire or cable we use in a solar system is very important. The distance between the solar inverter and the solar panels determines how thick the cable connecting the panels and the solar inverter should be. The greater the distance, the thicker the cable.

Since the electricity generated by the panels is of DC type, its voltage is low but the current is high. If the distance from the solar inverter to the panels is greater, the DC voltage generated by the panels decreases due to the resistance of the wire or cable and the expected DC voltage is not obtained at the connection points of the solar inverter. To solve this problem, the thickness of the wire or cable has to be determined according to its current and distance. We will take a detailed look at this in the next chapter.

Apart from wires and cables, various types of connectors are used to connect solar panels. Their types are MC4, sunclix or Y type. All these types of connectors also have male and female types. Solar panels come with 2 connectors. One is connected to the +ve terminal of the solar panel and one is connected to the -ve terminal. Through these connectors, we connect the panels in series or parallel.

To connect the panels in series, we use the same connectors that are present on the panels. When the +ve terminal of one panel is connected to the -ve terminal of another panel, those panels are connected in series. See the figure below.

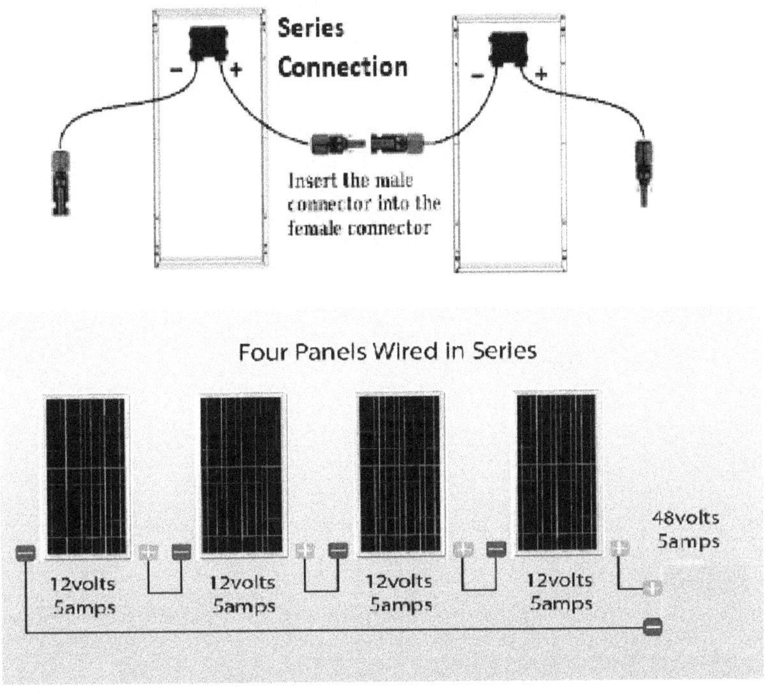

Series Connection of Panels

To connect panels in parallel, you need a Y connector. The +ve terminals of both panels are connected together and the -ve terminals are connected together.

See the parallel connection and connectors in the figure below.

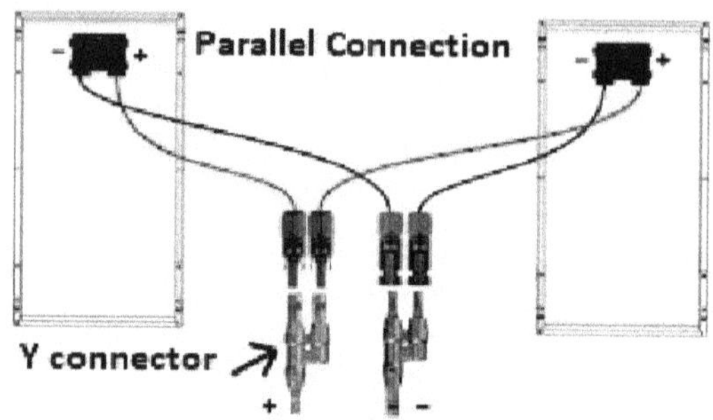

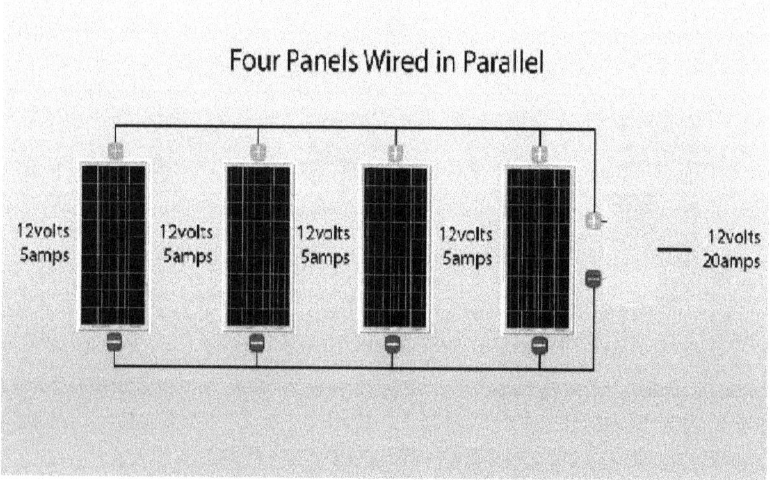

Parallel Connectio of Panels

There is DC wiring from the panels to the solar inverter and there is AC output wiring from the solar inverter to the load. DC and AC types of wires are different. We will learn more about that in the next section.

DESIGN OF THE SOLAR ENERGY SYSTEMS

In this very important part, let us see what type of solar system is more suitable for us, how much capacity of the system will be suitable and how we can design it ourselves. The entire process of which devices, in fact, what capacity of devices, and how to connect them to generate the solar energy we want is called design. The design of our system will depend on these 3 things: our energy requirement, load shedding times and how long the load shedding is. Let us take an example and design the complete system on paper so that we can understand the entire process.

Before designing the system, you need to note the following things -

1) What is your total electrical load? - The answer to this question tells you how much capacity your solar inverter should have.

2) Do you have frequent power outages? The answer to this question tells you what type of system you want, i.e. should you use an on-grid or hybrid system.

3) Do you have any essential load after sunset? And how long after sunset do we have load shedding? The answers to both questions tell us what the capacity of the battery should be in our system.

Design Steps -

1. How much capacity of the solar inverter should be taken.

2. Determine whether the system is on-grid or hybrid.

3. How much capacity and how many solar panels should be taken.

4. If it is a hybrid system, determine the capacity and number of batteries

5.. Determine the design of the structure by looking at the space available for installing the solar panels.

6. Determine the thickness of the AC and DC distribution boxes, wires and connectors.

7. Prepare the wiring diagram and connect and install accordingly.

Example -

Let us take an example and design a complete system. Let us use the table below to see what the electrical load is in our home, office or factory.

LOAD CHART		
Sr No	**Appliance**	**Watts**
1	Tube Light	60
2	LED light	20
3	Bulb	60
4	Night Lamp	5
5	Outdoor Light	100
6	Ceiling Fan	60
7	Exhaust Fan	60
8	Table Fan	50
9	TV	200
10	Fridge	400
11	Washing Machine	300
12	Air Conditioner	1500
13	Geyser	3000
14	Electric Iron	500
15	Mixer	100
16	OTG Oven	1000
17	Micro Wave	200
18	water pump	750
19	Vacuum Cleaner	800
20	Computer	250
21	Laser Printer	200

Let's say I want to install a solar power plant for both my office and my home. Let's design the system for both these plants. Let's assume that there is load at home even in the evening, so let's design a hybrid solar plant for the home and an on-grid solar plant for the office.

Solar Plant 1 Design - Home Solar Plant Design

Step 1 – Decide on the System, On Grid or Hybrid

I have frequent power outages. Since I have frequent power outages, an on grid solar system is not suitable for me. Since this solar plant is to be installed at home, there will be some load even in the evening, so a hybrid system is the right option.

Step 2 – How much capacity solar inverter to get

What is your total electrical load? - The total load in my house is as follows.

LOAD CHART				
Sr No	Appliance	Watts	Quantity	Tot Watts
1	LED light	20	6	120
2	Ceiling Fan	60	4	240
3	Exhaust Fan	60	1	60
4	TV	200	1	200
5	Fridge	400	1	400
6	Washing Machine	300	1	300
7	Air Conditioner	1500	1	1500
8	Mixer	100	1	100
9	water pump	750	1	750
10	Computer	250	1	250
11	Printer	200	1	200
	TOTAL WATTS			4180

The total load is 4180 watts, which means I need to use a 5 kw solar inverter. Note that we have considered a 1 phase inverter here. Some houses may have a 3 phase connection. If there is a 3 phase connection, the inverter will also have to be 3 phase and so will other components.

That is, we will be using a 5 kw single phase hybrid inverter.

In our example, we are using a single phase 5 kw inverter from the company Flynn Technologies. Let's look at its data sheet so that we can see how many panels and how many volts of DC the inverter operates on. This is a very important step because the inverter is the heart of the entire system, so it is very important to read its data sheet carefully.

I am giving the necessary information from the data sheet of Flin's inverter below.

1. Open circuit Voltage (Voc) of PV modules not exceeds max. PV array open circuit voltage of inverter.
2. Open circuit Voltage (Voc) of PV modules should be higher than min. battery voltage.

Solar Charging Mode				
INVERTER MODEL	1KVA 24V 2KVA 24V 3KVA 24V	1KVA 48V 3KVA 48V	2KVA 24V Plus/ 3KVA 24V Plus	2KVA 48V Plus/3KVA 48V Plus/ 4KVA/5KVA
Max. PV Array Open Circuit Voltage	75Vdc max	102Vdc max	145Vdc	
PV Array MPPT Voltage Range	30~66Vdc	60~88Vdc	30~115Vdc	60~115Vdc
Min. battery voltage for PV charge	17Vdc	34Vdc	17Vdc	34Vdc

Information from Hybrid Inverter Manual

Let us use the above information in our design context. Our inverter is 5 KVA / KW, so let's look at the last column in the table. In the first row, the inverter model is 5 KVA / KW in the last column and 5 KVA is written as 48 V. This means that this inverter requires 48V i.e. 4 batteries. These batteries will have to be connected in series to give 48 V to the inverter. The DC voltage coming from the solar panels is also there and its range is given in the last column in the second row which is 145 V. Now we have to connect our solar panels in such a way that we get a total of 145 volts by combining the series parallel connection. In the next step 3, we will decide how many solar panels we have, so we will look at the connection combination. In step 7, when we draw the wiring diagram, this series parallel connection will be clear.

Step 3 - How much capacity of solar panels and how much to take

Since it is a 5 kw solar inverter, we must also take solar panels of at least 5 kw. Inverters of some companies can also connect more solar panels than their own capacity. That is, inverters of some companies, even though they are of 5 kw capacity, connect 6 kw solar panels and generate more electricity to that extent. Let's design so that we connect solar panels as much as the capacity of the inverter. If one day all the loads continue to run throughout the day, the energy generated by all the panels will not be left to be used for battery charging, but sometimes it happens that all the loads (let's call them connected loads) will be running throughout the day. We are designing with the same number of panels as the load and as written above, we can use a little more panels.

For 5000 watts, for 330 watts,

5000/330 = 15.15 panels.

If we round the resulting number, it becomes 15, which means the number of solar panels is 15. These 15 solar panels have to be connected in series or parallel connection depending on the company of the inverter. We will see that in the next chapter.

Step 4 - If you have a hybrid system, determine the capacity and number of batteries

Batteries are usually available in 12 V. And their number depends on the type of inverter. That is, hybrid inverters operate on 12 V / 24 V / 48 V / 96 V or 192 V. This means, inverters operate on 1 battery / 2 / 4 / 8 or 16 batteries. Now the question remains, what should be the capacity of the battery, i.e. how many AH (ampere hours) should it be?

In the evening, after sunset, you have to determine the battery capacity based on how many hours you want to run the load. That is, to determine how much capacity of the battery you will need to run the required load for that number of hours. Let us do this calculation in our example.

In our example, we have taken the load of the house, out of which the load shown in the table below, you want to run the load shown in the table for the hours shown in the table after sunset.

LOAD CHART

Sr No	Appliance	Watts	Qty	Tot Watts	evening hours working	Tot Watt X Hours. वॅट्स X तास
1	LED light	20	6	120	5	600
2	Ceiling Fan	60	4	240	3	720
3	Exhaust Fan	60	1	60	0	0
4	TV	200	1	200	2	400
5	Fridge	400	1	400	4	1600
6	Washing Machine	300	1	300	0	0
7	Air Conditioner	1500	1	1500	0	0
8	Mixer	100	1	100	0	0
9	water pump	750	1	750	0	0
10	Computer	250	1	250	4	1000
11	Printer	200	1	200	0	0
	TOTAL WATTS			4180		4320

Enter Caption

The following formula is used to calculate the capacity of the battery.

(Watts X Hours) 4320

Battery AH = ------------------------ = --------- = 150 AH

0.6 X System Voltage 0.6 X 48

0.6 is a constant. This constant has been determined by considering the efficiency of the battery (level of discharge) and the efficiency of the inverter.

In our example, since the inverter system is 48 V, 4 batteries of 12 V will have to be connected in series and as per the above calculation, the AH of each battery is 150, so in total we will need 4 batteries of 150 AH. These batteries will supply current to the load given in the table above in the evening and this load will run for the required period.

To find out Battery sizing, use following formula

(Watts X hours) 4320

Battery AH = ------------------------ = 150 AH

(0.6 X System Voltage) 0.6 X 48

Step 5 - Determine the design of the structure by looking at the space available for installing solar panels

This solar plant is to be installed on the roof of the house and placed on the ground above the roof because usually the roof is waterproofed. The structure of the solar panels is such that it can be placed on the ground so that there is no need to do any drilling or demolition work on the roof. See the picture below.

Ground Mounted Structure- No drilling on roof is required

Structure with Solar Panels Installed

Structure customised in Rural Area

If ample space is available then such structture can be used

ON TIN Roof - This type of Structure is used

The angle of this structure should be equal to the longitude of the place where we are installing this solar plant. For example, in Mumbai, this angle should be 20 degrees because the longitude and latitude of Mumbai are 19.0760° N, 72.8777° E.

Generally, each solar panel is 2 meters X 1 meter in size. So, suppose we are installing 10 panels, we will need 12 meters X 3 meters of space. Or ..

Step 6 - Determining AC and DC Distribution Box, Wire Thickness and Connectors

AC Distribution Box-

In any solar system, 2 MCBs are installed on the AC side. One in the input AC line and one in the output AC line of the system. The first MCB can turn off the input power while the second can turn off the power given to the load and both these MCBs are required. Both these MCBs are required because they act as a protective device of the system during any maintenance or repair work. If any failure or fault occurs, the MCB trips and separates that circuit from the system.

There is no need to install a surge protection device (SPD) on the AC output side but it is useful to install a surge protection device (SPD) on the AC input side because surges can come on the line from outside. Surge means that the voltage increases a lot for a very short time and then comes back down, which is called surge. Surge protection devices are necessary as surges can damage parts or components in a solar system.

The AC distribution boards are available ready-made as shown below, and need to be brought and installed.

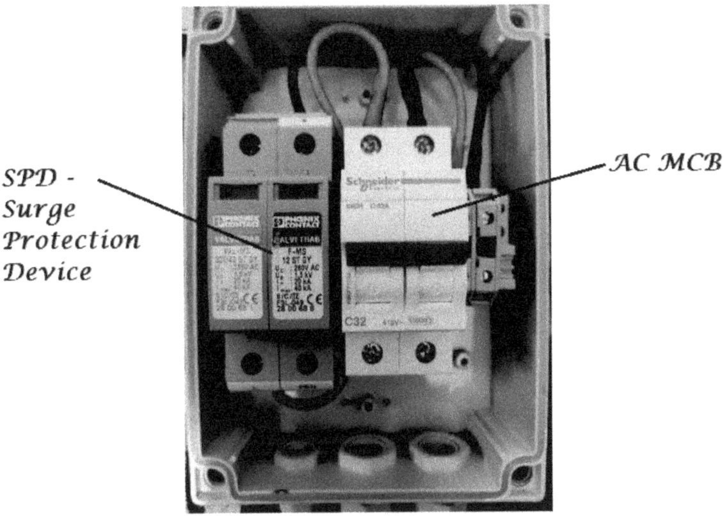

Single Phase AC DB with Surge Protection Device (SPD) for 1-5 KW
solar plant

DC Distribution Box-

In any solar system, 2 MCBs are also installed on the DC side. One in
the wires coming from the solar panels and one in the battery circuit of
the system. The first MCB can turn off the power coming from the solar
panels while the second can turn off the power given for battery charging
and both these MCBs are necessary because they act as a protective device
of the system during any maintenance or repair work. If any failure or fault
occurs, the MCB trips and separates that circuit from the system. That is, if a
failure occurs in the solar panels circuit, the MCB trips and disconnects the
solar panels from the system.

It is not necessary to install a surge protection device (SPD) in the
battery circuit, but it is useful, if not necessary, to install a surge protection
device (SPD) in the wires coming from the solar panels, because if there is a
lightning strike outside during the rainy season or if an electric pole passes
nearby, a surge from the solar panels can accidentally enter the circuit.

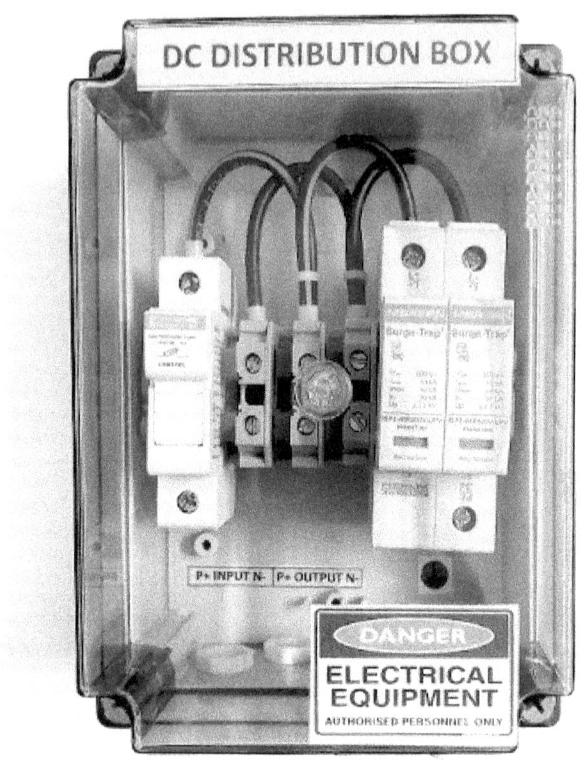

Single Phase DC DB with FUSE & Surge Protection Device (SPD) for 1-5 KW solar plant

AC Wire or Cables

Below Table is for wire or cable gauge or thickness

Cable size in mm²	Cable thickness in mm	Amp rating of cable
0.5mm²	1.7mm	11A
1mm²	2.1mm	16.5A
1.5mm²	2.3mm	21A
2mm²	2.8mm	25A
3mm²	3.4mm	33A
4mm²	3.8mm	39A
6mm²	4.3mm	50A
8.5mm²	5.6mm	63A
10mm²	6mm	70A
16mm²	8.3mm	110A
25mm²	10.1mm	170A
35mm²	11.8mm	240A
50mm²	13.3mm	345A
70mm²	15.5mm	485A
95mm²	17.9mm	500A

AC Wires or cables thickness

Inverter Capcity	Gauge
1KVA	16 AWG
2KVA 230VAC	14 AWG
2KVA 120VAC 3KVA	12 AWG
4KVA	10 AWG
5KVA	8 AWG

Inverter size and wire gauge

AWG (American Wire Gauge) American Wire Gauge. This is an American standard. The larger the number, the thinner the wire, and the smaller the number, the thicker the wire.

DC Wires -

When connecting battery wires to the battery terminals, the ring terminal connectors or similar terminal connectors given below must be used. The picture of the ring terminal is given below and the table below it gives the battery wires as well as the sizes of these ring terminals.

Ring terminal:

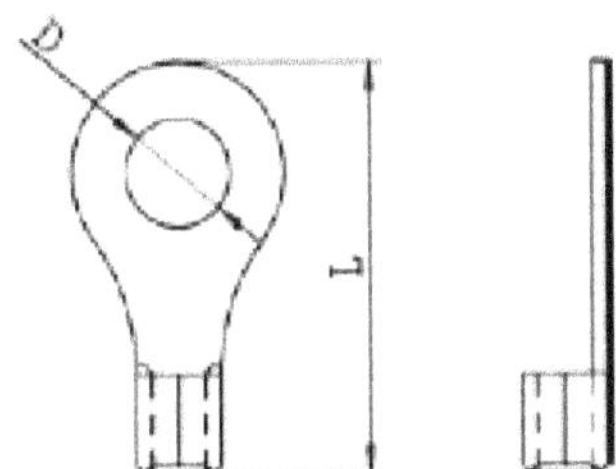

DC cables need this type of terminals to fix wires or cables as they are bit thick

Inverter Capacity	Typical Amperage	Battery Capacity	Wire Size	Ring Terminal		
				Cable	Dimensions	
				mm^2	D (mm)	L (mm)
1KVA 48V	20A	100AH	1*14AWG	2	6.4	21.8
1KVA 24V, 2KVA 48V	33A	100AH	1*10AWG	5	6.4	22.5
3KVA 48V	50A	100AH	1*8AWG	8	6.4	23.8
2KVA 24V	66A	100AH	1*6AWG	14	6.4	29.2
		200AH	2*10AWG	8	6.4	23.8
3KVA 24V	100A	100AH	1*4AWG	22	6.4	33.2
		200AH	2*8AWG	14	6.4	29.2
4KVA	66A	200AH	1*4AWG	22	6.4	33.2
			2*8AWG	14	6.4	29.2
5KVA	87A	200AH	1*4AWG	22	6.4	33.2
			2*8AWG	14	6.4	29.2

DC wires cable size chart

In the above table, the inverter capacity is shown in KVA instead of KW. We can also use this table for the wires coming from solar panels. We should calculate the current and determine the thickness of the wire as we connect solar panels in series and parallel. Generally, parallel connection is not required up to 10KW. For information, it is necessary to mention here that if parallel connection is made, the current in both the parallel wires should be added. No matter how many panels are connected in series, the current remains the same, that is, for example, if 10 panels of 360 Wp are connected in series, their current remains 10 Amp. But suppose 5 panels of 360 Wp are connected in series and 5 more panels are connected in series and their parallel connection is made, then 10 Amp + 10 Amp will be obtained from the panels, which will be 20 Amp current. Calculate the current in the above way and determine the thickness of the wire. See both the pictures below.

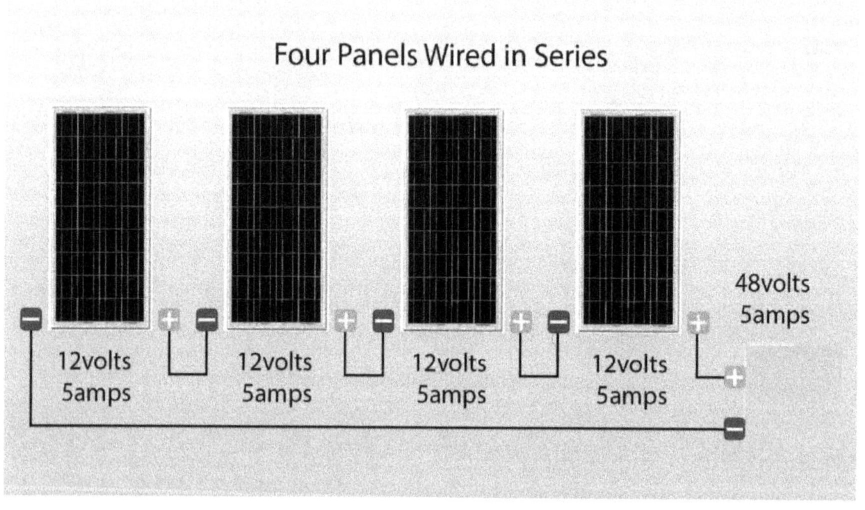

In series connection - Voltage gets added and Current remains same

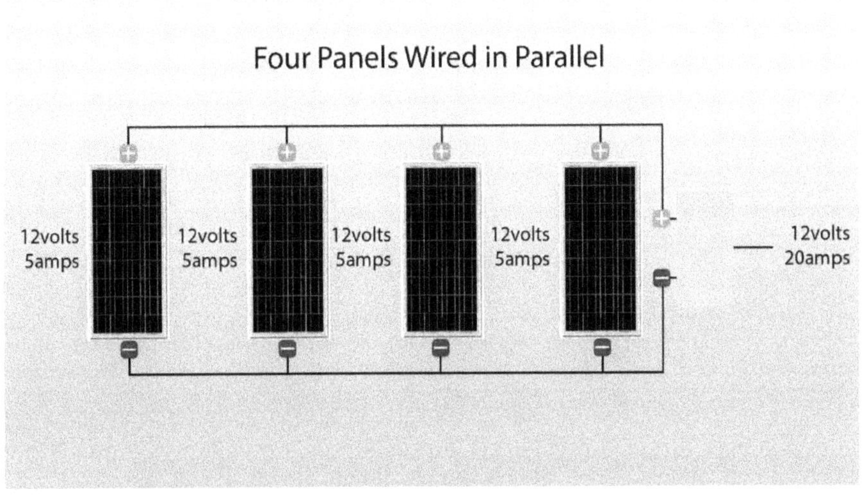

In Parallel Connection - Voltage remains same and Current gets added

Connectors –

It is very easy to make a series connection of solar panels because there are 2 connectors for connecting solar panels. One is connected to the +ve terminal of the solar panel and one is connected to the -ve terminal. Through these connectors, we connect the panels in series or parallel. To connect the panels in series, we use the connectors on the panels. The panel has 2 connectors, +ve male and -ve female. When the +ve terminal of one panel is connected to the -ve terminal of the other panel, the panels are connected in series. See the figure below.

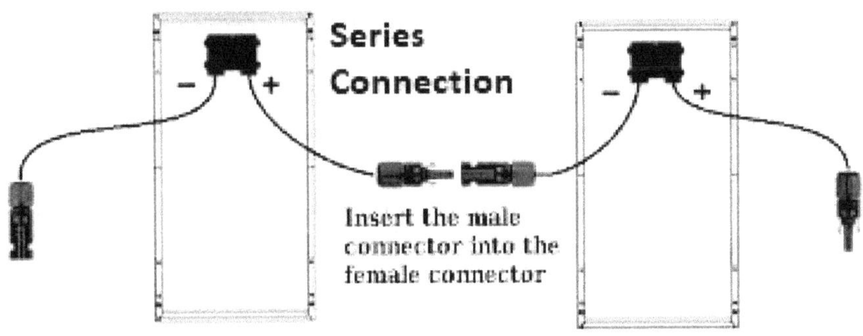

Series Connection of panels with the help of MC4 connectors

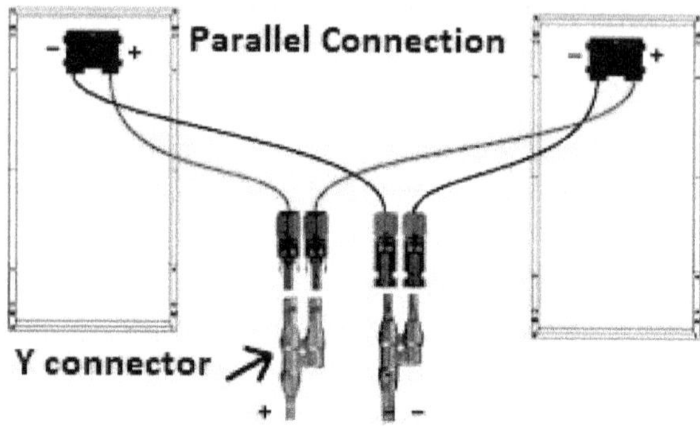

Parallel Connection of panels with the help of Y connector and MC4 connectors

Generally, all inverter and panel manufacturers use a standard, i.e., the same type of connectors. These connectors are called MC4 type connectors. Some inverter manufacturers, like SMA, use a different type of connector, which is called Sunclix connectors.

Step 7. Prepare the wiring diagram and connect and install accordingly.

In the above 6 steps, we have come to this design that
• The system is of hybrid type
• Inverter capacity is 5 KW
• 15 solar panels of 330Wp capacity
• 4 batteries of 150 AH, 12 V capacity
• Ground structure
• AC distribution box, DC distribution box

The maximum DC voltage of our inverter is 145 V and the MPPT (Maximum Power Point Tracker) voltage range is 60 - 115 Volts. Our panels are of 330 Wp. The voltage of each of those panels is 36 volts. To keep the voltage of the panels in this MPPT (Maximum Power Point Tracker) voltage range, you will have to connect 3 panels in series so that their voltage becomes 36 X 3 = 108 volts, i.e. it remains in the MPPT range. The total number of panels is 15, so you will have to connect 3 panels in series and 5 sets of 3 panels in parallel.

MPPT - means the voltage range of the DC voltage coming from the panel in which the inverter delivers maximum power.

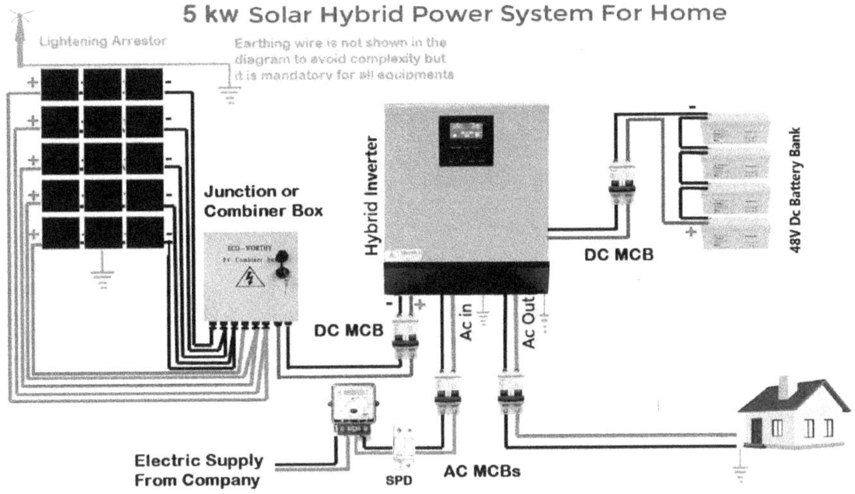

Complete wiring diagram of a hybrid 5KW solar System

In the wiring diagram above, both the DC and AC MCBs are installed in the same box. We call that box the distribution box. Please note that only the distribution box is shown in the diagram to make the wiring easier to see. Also, the SPD i.e. surge protection device is also installed in the distribution box.

Design of Solar Plant 2 - Design of Solar Plant for Office

Design of Solar Plant for Office. In the design steps of the hybrid solar plant designed for the house, only step no. 4 - Battery design, is not required. All the remaining steps will remain the same, but let's see the design of the on-grid solar plant step by step.

1 - Deciding whether the system is on-grid or hybrid

The office is open during the day and there is probably no load in the evening, so the office is installed with an on-grid system. Let's also see the design of the on-grid system.

2 - How much capacity of solar inverter to take

What is your total electrical load? - The total load of my office is as follows.

LOAD CHART				
Sr No	Appliance	Watts	Quantity	Tot Watts
1	LED light	20	30	600
2	Ceiling Fan	60	10	600
3	Exhaust Fan	75	2	150
4	Fridge	400	1	400
5	Air Conditioner	1500	3	3000
6	Computer	250	20	250
7	Printer	200	4	200
8	Other Hardware	1000	lumpsum	1000
	TOTAL WATTS			6200

LOAD for office Example

The total load is 6200 watts, which means I need to use a 7 kw / 7.5 kw solar inverter. Note that we have considered a 1 phase inverter here. Some offices may have a 3 phase connection. If there is a 3 phase connection, the inverter will also have to be 3 phase and so will other components.

That is, we will be using a 7kw / 7.5kw single phase on grid inverter.

In our example, we are using a single phase 7.5 kw inverter from the company Sofer. Let's look at its data sheet so that we can see how many panels and how many volts of DC the inverter operates on. This is a very important step because the inverter is the heart of the entire system, so it is very important to read its data sheet carefully.

I am giving the necessary information from the data sheet of Sofar inverter below.

Datasheet — SOFAR 7.5KTLM

Input (DC)

Recommended Max. PV input power	9980Wp
Max DC power for single MPPT	2*2750W/ 3000W
Number of MPP trackers	
Number of DC inputs	2/1
Max. Input voltage	
Start-up voltage	
Rated input voltage	
MPPT operating voltage range	
Full power MPPT voltage range	250V-520V
Max. Input current per MPPT	22A/11A
Maxnimun DC input short circuit current per MPPT	26.4/13.2A

Output (AC)

Rated power	7500W
Max. AC power	7500VA
Max. Output current	32.6A
Nominal grid voltage	
Grid voltage range	
Nominal frequency	
Grid frequency range	
Active power adjustable range	
THDi	
Power factor	
Power limit export	

Performance

Max efficiency	98.2%
European weighted efficiency	97.6%
Self-consumption at night	
MPPT efficiency	

Protection

DC reverse polarity protection	Yes
DC switch	Optional
Protection class/overvoltage category	I/III
Safety protection	Anti islanding, RCMU, Ground fault monitoring
SPD	MOV: Type III standard

Communication

Power management unit	According to certification and request
Standard communication mode	RS485, Wifi/Ethernet/GPRS/4G(optional), SD card(optional)
Operation data storage	25 years

General Data

Ambient temperature range	-25°C ~ +60°C
Topology	Transformerless
Degree of protection	IP65
Allowable relative humidity range	0~100%
Max. Operating altitude	2000m
Noise	<25dB
Weight	11.5kg
Cooling	Natural
Dimension	405*315*135mm
Display	LCD display
Warranty	5 years/ 7 years/ 10 years

SOFAR Inverter Data Sheet as an example

Let us use the above information for our design. Our inverter is 7.5KW so let us look at the INPUT DC section in the table.

We need to note down the following information from the data sheet of this solar inverter. Later, we will need this information while preparing the wiring diagram.

INPUT DC Information

1. This inverter has 2 MPPT inputs, which means we can divide the solar panels into 2 strings and connect them.

2. The maximum MPPT voltage for this inverter can be 520 v.

OUTPUT AC Information

3. Output current is 32.6 Amp and accordingly you will have to install MCB of 63 AMP.

4 - How much capacity of solar panels and how much to take

During the hybrid inverter design, we saw that the solar panels are of the same KW as the load.

For 6200 watts,

63000/350 w = 18 panels.

Let us divide it into 1 string of 9 panels and 1 string of other 9 panels. These 18 solar panels have to be connected in series connection according to the company of the inverter. We will see it in the next chapter. Due to this division, we will also get the MPPT voltage. Each panel gives 36 volts, that is, a string of 9 panels is 36 x 9 = 324 volts. This is less than the maximum MPPT voltage.

In step 7, we will see the connection combination and draw the wiring diagram, when this series connection will be clear.

5 - If there is a hybrid system, determine the capacity and number of batteries

Since there are no batteries in this on-grid system, this step will not be in this design.

6 - Determine the design of the structure by looking at the space available for installing solar panels

Installing solar panels is a mechanical work and the system is the same regardless of the manufacturer, so this step should be understood as step 5 of the hybrid system design.

7 - Determine the AC and DC distribution box, thickness of wires and connectors

In any solar system, 2 MCBs are installed on the AC side. One in the input AC line and one in the output AC line of the system. The first MCB can switch off the input power while the second can switch off the power given to the load and both these MCBs are required. Both these MCBs are necessary because they act as a protective device of the system during any maintenance or repair work. If any failure or fault occurs, the MCB trips and separates that circuit from the system. There is no need to install a surge protection device (SPD) on the AC output side, but it is useful to install a surge protection device (SPD) on the AC input side because a surge can come on the line from outside. Surge means that the voltage increases a lot for a very short time and then comes back down, which is called a surge. Surge can damage the parts or components in the solar system, so a surge protection device is necessary.

This step is also very similar in hybrid and on-grid systems, except that there is no DC MCB for the battery in the distribution box. The rest of the complete setup is the same for hybrid and on-grid.

8. Prepare the wiring diagram and connect it accordingly and do the installation.

In the above 6 steps, we have come to this design that

- The system is of on-grid type
- Inverter capacity is 7.5 KW
- 18 solar panels of 350Wp capacity
- Ground structure
- AC distribution box, DC distribution box

The maximum DC voltage of our inverter is 600 V and the MPPT (Maximum Power Point Tracker) voltage range is 250 - 520 Volts. Our panels are of 350 Wp. The voltage of each of those panels is 36 volts. To keep the voltage of the panels in this MPPT (Maximum Power Point Tracker) voltage range, we have to connect 9 panels in series so that their voltage becomes 36 X 9 = 324 volts, that is, it remains in the MPPT range. The total number of panels is 18, so we have to connect 9 panels in series and 2 sets of such 9 panels to 2 separate DC inputs in the inverter.

MPPT - means the voltage range of the DC voltage coming from the panel in which the inverter delivers maximum power.

Now let's look at the wiring diagram of our design so that we can move on to the next part of the installation.

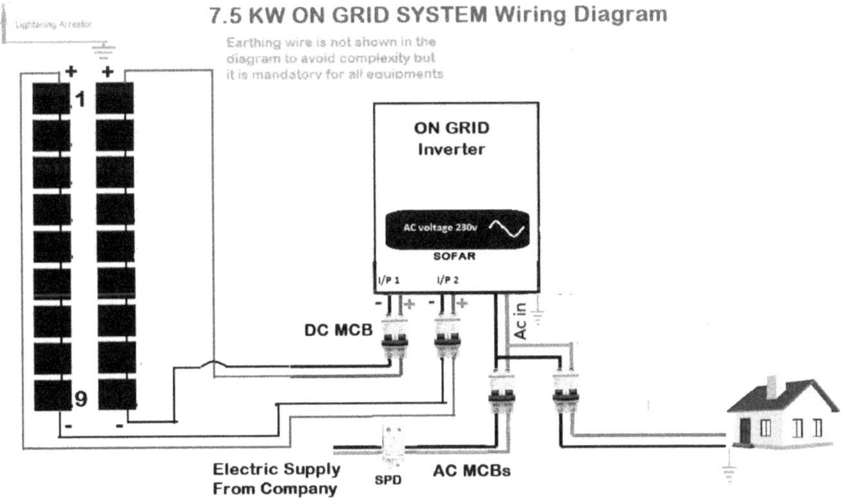

Complete Wiring diagram of ON GRID Solar system as an example for Office

INSTALLATION OF THE SOLAR ENERGY SYSTEM

Chapter 8 – INSTALLATION OF THE SOLAR ENERGY SYSTEMS

1. Erection of Structure

As we have seen in the previous chapters, out of the different structures, you have to install the structure that is convenient for you. We install the panels on the structure, so you should check if the angle of the solar panels is according to the longitude of your place. For example, if your installation is in Mumbai, this angle is 18 to 20 degrees.

2. Fitting Strut Purlins to Install Solar Panels on the Structure

No matter what your structure is, you need to fit strut purlins on the structure to install these solar panels. You also need to use strut purlins and special types of nuts and U or Z type clips. See the picture below. In this way,

install all the panels on the structure using strut purlins, nuts and suitable clips.

Structure for 2 panels

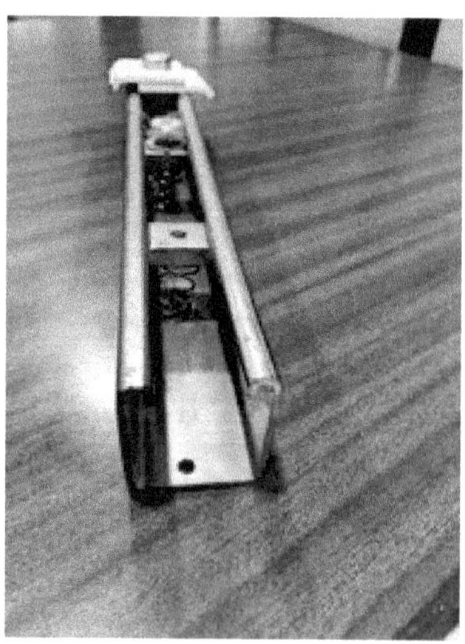

Special NUT BOLTS to be used with Strut Channels

Solar Panels installed with Channel and Clamps, NUT BOLTS

3. Mounting Solar Panels

The same strut purlins, nuts or channels and special nuts are used to mount solar panels on different structures. See the pictures below. The pictures clearly show how the panels are mounted.

Mounting with various tyoes of structures

4. Connecting the wires of the panels

A wiring diagram is prepared as per the design and the wires of each panel are connected accordingly. As many panels should be connected in series and as many in parallel as possible. As we have seen in the previous chapters, each panel is connected with connectors called MC4. They can be connected to each other without any screw driver or pliers. The positive and negative terminals fit and connect to each other by simply pressing them.

In this way, the panels can be connected by simply connecting the connectors to each other in series parallel. After connecting all the panels, there should be 2 wires left at the end, one red and one black, positive and negative.

5. Installing Solar Inverter

Inverters come from many companies, for example SMA, Sofer, Schneider, Flynn and others. We saw the Flynn company inverter in the design example here. Let us see the inverter of the same company in this installation case.

Remove the inverter from the packing box and keep it in a safe place. Keep the information sheet, wires, connectors, CD that comes with it, because these items can be used later. You have to install it as per the pictures given below. All the pictures are clear and guide.

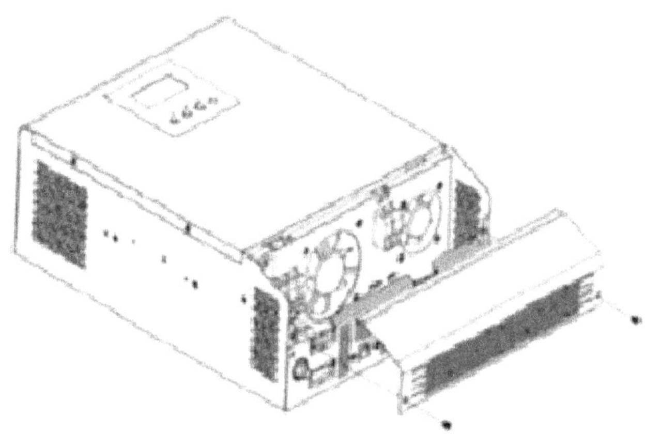

Opening the Inverter for connections

Remove the 2 screws and remove the bottom cover, which means that all the connection screws are visible.

The inverter should be fixed on a dry wall in the following way. There should be nothing else between 20 and 50 cm on its right side as shown in the figure. With the help of 3 screws, the inverter should be fixed on the wall at the level of your face because the indicator or display on it will be easy for you to see.

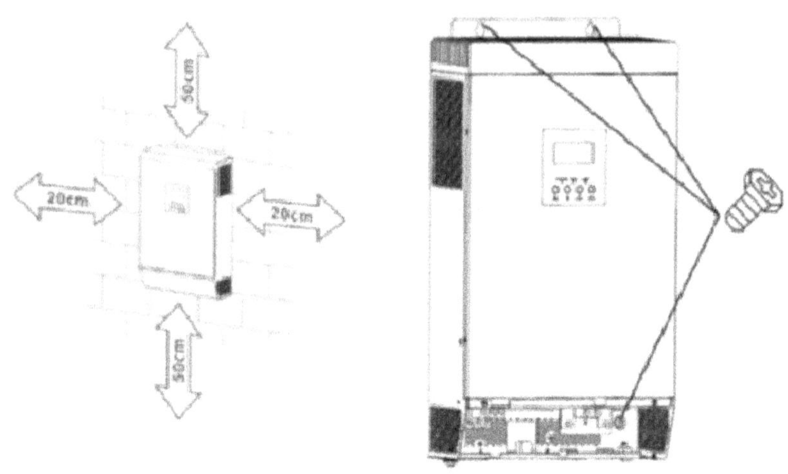

Inverter fixing on Wall

6. If there is a solar panel, inverter and hybrid system, also connect the battery to each other

The solar panel connection should be done as shown in the figure below and the connection should be done properly by observing +ve and -ve

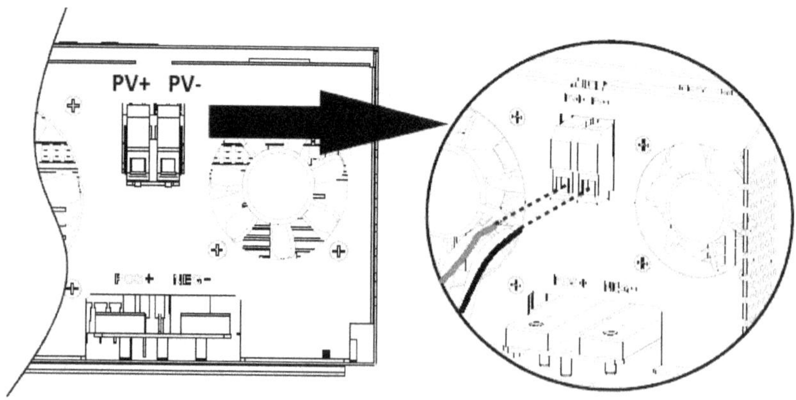

Solar Panels Connection to Inverter

Picture of connecting the battery. The battery has 2 different connections namely +ve and -ve, connect the wires by looking at them. If the inverter system is of 12 volts, then 1 battery, if it is of 24 volts, then 2 batteries and if it is of 48 volts, then 4 batteries have to be connected in series. We have seen battery connections in the chapter on batteries.

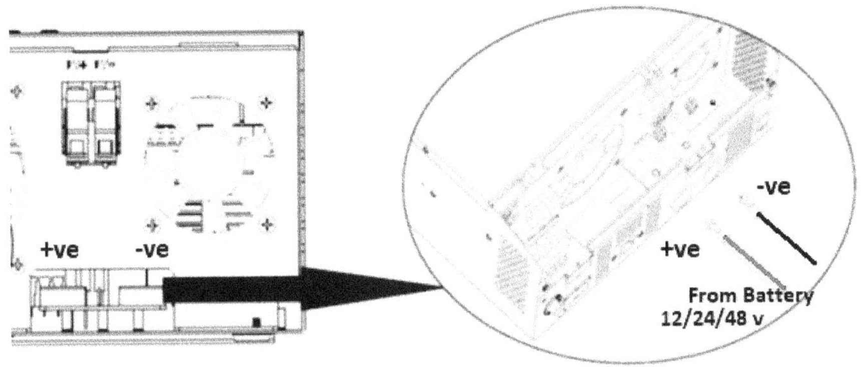

Battery Connection to Inverter

AC Input and Output Connections is to be done as shown below

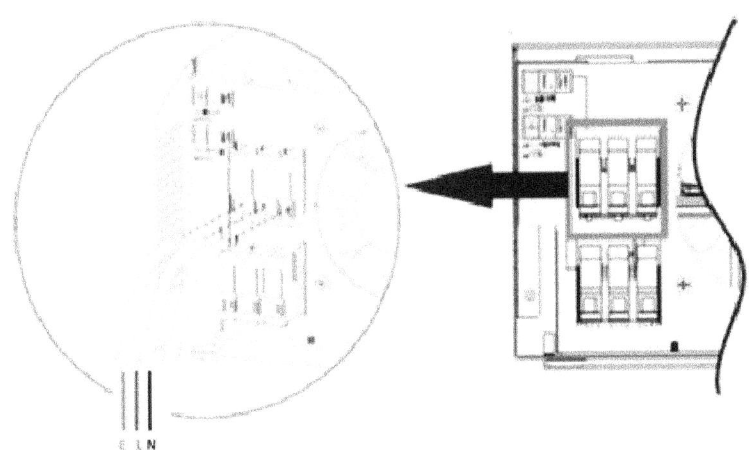

AC INPUT Connection

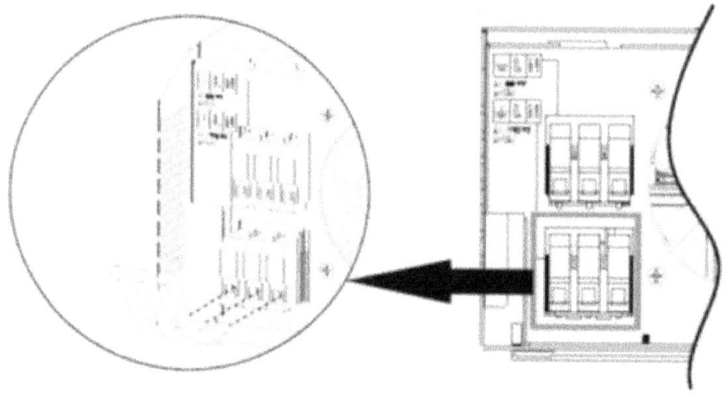

AC OUTPUT Connection

7. Connect the inverter to the grid supply

Connect the wires coming from the AC MCB in the AC distribution box to the inverter AC input

8. Recheck all the connections

Note that all the wires of the panels, batteries, AC input and output should come from the distribution box. See the picture below. In this way, all the connections should be made from the distribution box.

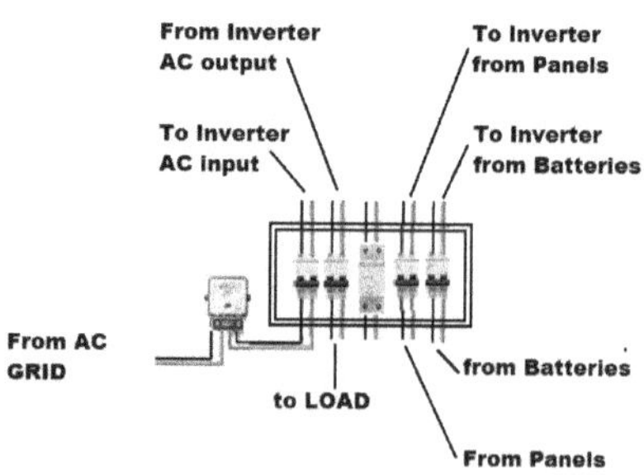

Rechecking all the connections

As shown in below figure, post checking final connections of wires, cover is fixed with the help of screws.

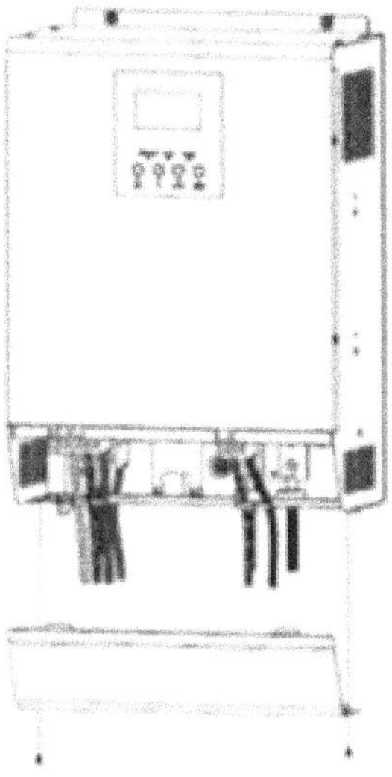

All Connections done- Panels, Batteries, AC Input, AC Output

9. Turning on the inverter

As shown in the wiring diagram, turn on the AC input MCB and supply AC to the inverter. Then turn on the battery MCB and connect the batteries to the inverter. The inverter will not start unless the batteries are connected. If no fault indicator appears after that, turn on the panel MCB and connect the panels to the inverter. Now the inverter is fully turned on.

After the inverter gets AC supply, it usually takes 3 minutes to wait because the inverter checks all the settings and then starts generating electricity. The status of the inverter is known on the front display of the inverter. Generally, all inverters have indicators for input voltage, voltage coming from the panels and batteries. Some inverters also display more information than this. If there is any problem, it appears on the display.

10. Turning on and checking the load

First of all, make sure that the load is not more than the capacity of the inverter. Because if the load is more than the inverter capacity, the inverter will shut down by giving an overload signal. Increase the load one by one and finally see if the inverter is taking all the load or not. In this way, put the full load on the inverter and check and make sure.

After putting the full load on the inverter, it will be useful to check whether your load runs as designed or for as long as the design of your system. If the expected performance is not getting, what changes need to be made in the system can be considered.

If the solar energy is less for the load during the day, you can consider increasing the panels or if the load does not run for the expected time after sunset, you can consider increasing the battery capacity.

CHAPTER NINE

COST & MAINTENANCE OF SOLAR ENERGY SYSTEM

9 - COST & MAINTENANCE OF SOLAR ENERGY SYSTEM

Cost of Solar System -

The cost of solar system is generally calculated per watt. While calculating it, the following points are also calculated.

1. Inverter - Rs 15/- per watt for 1 phase & Rs 25/- per watt for 3 phase inverter

2. Panels - Rs 25/- per watt

3. Structure - Rs 10/- per watt can also be reduced according to location

4. Wiring and Installation - Rs 15/- per watt

TOTAL/watt - Rs.65/- per watt 1 phase

The battery size is determined by the number of hours the load is to be run in the absence of sun or solar energy, so the cost of the battery has to be directly added to the above total per watt cost. And then the total cost of the solar system will be obtained.

Materials required for solar repair and maintenance:

Catalog or information sheet of solar panels, inverter, battery, distribution box. AC/DC wire size chart, screwdriver, pliers, MC4 connectors, WD40 spray, multi meter, tester, PVC tape and other materials that may be applicable on site

Inverter -

While maintaining the inverter, the DC MCB of the panel, battery and the MCB of the AC input output should be turned off from the distribution box. This will turn off the inverter. Remove the cover of the inverter.

Then the connection terminals of the inverter should be cleaned. Then the MCB of the panels, battery, AC input and output should be turned on in this order. Measure the panel, battery, AC input and output voltage on the inverter connection terminals. Check whether the voltage is coming correctly on those terminals according to the number of panels and batteries. If the AC input output gets 230 volts, then we can conclude that the inverter is working properly. If the correct voltages are not coming, contact the inverter company representative.

Solar Panel -

Measure the Voc of the solar panel, i.e. the open circuit voltage, with a multimeter. If it is very different from what is shown on the plate of the solar panel or in its catalog, then the solar panel is faulty. There is a diode in the terminals of the solar panels. If the Voc value shows 0, then it is more likely that this diode is gone. In that case, the diode needs to be replaced. After doing this, measure the voltage again and if correct, put the panel back into the circuit.

Battery -

First of all, measure the open circuit voltage of the battery. That is, when none of the panels, supply or load is connected to the battery, the voltage reading we take is called the open circuit voltage. This Voc of a fully charged battery should be 12.54 v. If it is less than that, then start further trouble shooting. If the battery voltage drops below 11 v, then it is considered that the battery is discharged.

The battery should always be kept away from ventilated and flammable materials because hydrogen is released from it. This hydrogen causes a white substance (PbSo4) to accumulate on its terminals. To avoid this, the terminals should be kept clean and greased or petroleum jelly should be applied to them regularly. The level of the battery's water i.e. electrolyte should be maintained at all times and the specific gravity of the sulfuric acid in the water should be measured with a meter. If it is found to be low, only battery grade acid should be added to the battery. Do not use any other grade.

The specific gravity of the water or acid of a fully charged battery is measured to determine whether the battery is fully charged. It is measured with a specific gravity meter. The specific gravity of the battery acid should be 1.250. If it is found to be below 1.2, it should be charged and if the specific gravity does not increase even after doing so, it should be considered that the battery is damaged. The meter shown below should be used to measure the specific gravity. It is called a hydrometer.

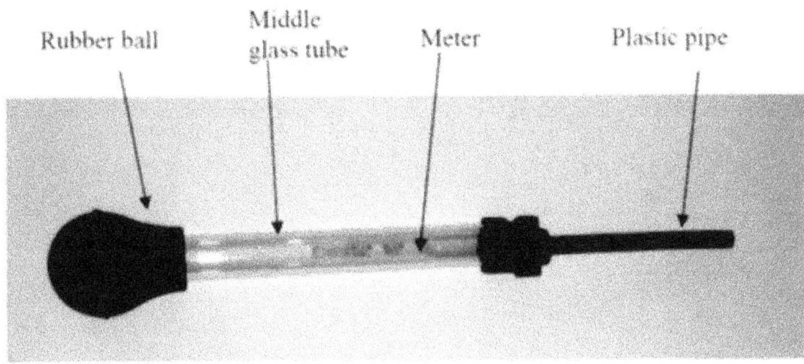

Device to check the BATTERY

By leaving its open tube in the battery acid, by pressing the rubber balloon and slowly releasing it, the acid should be completely filled in the glass tube and then by releasing the pressure on the rubber balloon, the reading on the glass tube should be taken. This is the reading of the specific gravity of the battery acid which should not be less than 1.250 for a fully charged battery. If the battery is charged in this way but the specific gravity of its acid does not show 1.250, then it should be considered that there is a problem in the battery. It should be shown to a battery specialist or charger.

In this way, the maintenance of the entire solar system should be done regularly at regular intervals so that the solar system continues to give good output.

CHAPTER TEN

INNOVATIVE PRODUCTS

There are many new discoveries in the field of solar energy. Not just discoveries, but also products suitable for daily use are coming to the market. I would like to focus on some products that can enrich our lives.

1) As seen in the picture, we can easily install thin film panels on the roof of the house without using any heavy structure to generate solar energy.

Thin Film Solar Panels - Stuck on roof, no structure is required & light in Weight

THIN FILM Solar Panels installed

2) The panels available now cannot absorb ultraviolet (infrared) rays, but now the new panels, which have small carbon wires, generate more energy by using ultraviolet (infrared) rays in the sun's rays. Since such panels are transparent, we can use them on windows, mobile screen guards or similar devices. In the near future, such panels can be used to get rid of mobile charging by using mobile covers or screen guards. By applying such panels to the glass of windows, we can run the common lights of the building on them.

Transperent Solar Panels fro Facade

3) It has now been proven that paints made from lithium ion polymers can be used as solar energy sources. If you apply such paints to the roof or walls of your house, they will store energy in the paint during the day and glow in the dark. General Electric Company has developed such paints which will soon be available in the market. They have been developed at the National Renewable Energy Lab, Colorado, USA.

4) Jain Irrigation Systems and MIT, USA have jointly developed a plant to make seawater drinkable which runs entirely on solar energy. This does not require any other energy and can also solve the problem of water.

Desalination of water based on solar power

5) In a European country, instead of building a concrete road, a road has been built with solar panels, which allows solar energy to be generated throughout the day.

ROAD made out of solar panels

We should not be surprised if products based on such discoveries soon come to the market. Solar energy holds a great future for developing countries like ours, which can meet our growing energy needs while maintaining environmental balance. In the future, solar energy will become the fastest engine of our development.

CHAPTER ELEVEN

FUTURE

In our country, the sun has been worshipped as a life-giving deity since ancient times. Now in the era of industrial revolution, the sun has become a huge energy source. Sunlight is available in abundance in our country India. Nature has given our country so much sunlight that we can generate 4-5 units of energy per square meter of land. As mentioned, there is so much land in the country that we can generate 5000 crore units of solar energy every year. This energy is so much that the growing need of our entire country can be met by it. Going further, it can be said that this solar energy can be generated in a distributed form even in small villages. Also, the excess energy can be stored in batteries as per the requirement.

In this way, we can generate electricity even in very remote areas and make the villages there self-sufficient by providing electricity to the villagers. By connecting these distributed solar energy sources to each other, we can create a solar grid which can ensure energy security for every person, industry, factory etc. If everyone in the country gets energy security i.e. abundant energy, we can achieve sustainable development of the country which is environment friendly and pollution free. From the picture given below we can see the increasing use of solar energy. There is no alternative to solar energy in our country and I am sure that we will use solar energy as the largest energy source.

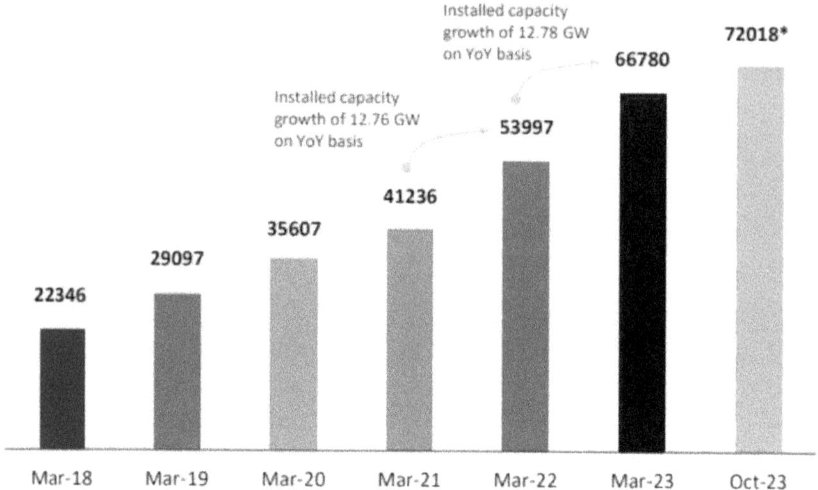

*Solar installed capacity in India as on 31st October 2023.

Increasing Generation of Solar Power

CHAPTER TWELVE

SCHEMES FOR SUBSIDY FOR SOLAR ENERGY

Government Subsidy Schemes for Solar Plant Installations. Each scheme targets different sectors such as residential, commercial, and agricultural users. Below, we'll explain the key schemes and their benefits.

1. PM Kusum Yojana (Pradhan Mantri Kisan Urja Suraksha evam Utthaan Mahabhiyan)

The **PM Kusum Yojana** is one of the most prominent solar subsidy programs, primarily designed for farmers. This scheme offers financial assistance for the installation of solar pumps and grid-connected solar plants. The initiative is intended to help farmers reduce their dependence on grid electricity or diesel-powered pumps, ultimately leading to lower energy costs.

Key Features:

- **60% Subsidy** for the installation of solar pumps and plants.
- **Component A**: Solar plants with a capacity of 500 kW to 2 MW that can be used to sell electricity to the grid.
- **Component B**: Installation of solar-powered standalone pumps for irrigation.
- **Component C**: Solarization of existing pumps (15 lakh pumps).

Benefits:

- **Affordable Irrigation**: Farmers can reduce their electricity or diesel consumption.

- **Increased Income**: Surplus energy can be sold to the grid, generating additional income.
- **Sustainable Farming**: Promotes cleaner, more energy-efficient farming practices.

Table: Subsidy Breakdown for PM Kusum Yojana

Component	Description	Subsidy Percentage	Beneficiaries
Component A	Solar plants for farmers (500 kW to 2 MW)	60%	Farmers, Agricultural Societies
Component B	Solar-powered standalone pumps	60%	Farmers
Component C	Solarizing existing pumps	60%	Farmers

2. PM Solar Yojana (PM Surya Ghar Muft Bijli Yojana)

Launched in 2024, the **PM Solar Yojana** aims to promote the installation of rooftop solar systems for residential properties. This scheme encourages homeowners to switch to solar energy by offering subsidies for rooftop solar panels.

Key Features:

- ₹30,000/kW **Subsidy** for up to 2 kW systems.
- ₹18,000 **extra** for 3kW systems up to 10kW.
  This sums up to a total of 78,000 for systems equal to one above 3kW.
- **Eligibility**: Available to Indian citizens owning residential properties with a roof suitable for solar panel installation.

Benefits:

- **Reduced Electricity Bills:** Solar power generated from the rooftop can help homeowners reduce their monthly electricity bills.
- **Net Metering:** Homeowners can sell surplus electricity back to the grid.
- **Eco-Friendly:** Contributes to reducing carbon emissions and reliance on fossil fuels.

Table: Subsidy for Residential Solar Systems

System Capacity	Subsidy Amount	Maximum Subsidy Amount	Annual Savings (Approx.)
Up to 2 kW	₹30,000/kW	₹ 60,000	₹20,000 to ₹30,000
3 kW to 10kW	₹60,000 + ₹18,000	₹ 78,000	>₹30,000

3. State-Specific Subsidies

In addition to national schemes, Uttar Pradesh also offers its own solar panel subsidy. Under the state solar subsidy provided by UPNEDA, the customers get a benefit of ₹15,000/kW up to ₹30,000 max. on solar rooftop installations.

4. Net Metering

Under the **Net Metering** system, individuals or businesses with rooftop solar panels can send the excess power generated back to the grid. This process helps users earn credits for the surplus energy produced. These credits can offset future electricity consumption, providing additional savings.

Key Features:

- **Reduction in Electricity Bills:** Excess solar energy can be sold back to the grid, leading to reduced monthly bills.
- **State-Level Variations:** Each state has its own net metering policy, so users must verify the rules applicable to their region.

How Does Solar Panel Subsidy Work?

The process of availing the **solar panel subsidy** is simple and involves a few key steps:

1. **Eligibility Check**: First, you need to verify if you meet the eligibility criteria for the subsidy scheme you are interested in.
2. **Selection of Vendor**: Choose a government-approved vendor who will handle the installation of your solar system.
3. **Installation**: Once the vendor is selected, the installation process will begin.
4. **Subsidy Disbursement**: The subsidy is either directly transferred to the vendor or applied to reduce the installation cost.
5. **Net Metering Setup**: If applicable, a net meter will be installed to track the energy produced and consumed.

Eligibility Criteria for Solar Panel Subsidy

The eligibility criteria for receiving a **solar panel subsidy** depend on the specific scheme. However, here are some common requirements:

- **Indian Citizenship**: The applicant must be an Indian citizen.
- **Property Ownership**: The property where the solar panels are to be installed must be owned by the applicant.
- **Roof Availability**: The property must have a roof that is suitable for solar panel installation.
- **Grid Connection**: The property must be connected to the electricity grid for schemes that involve net metering.

How to Apply for Solar Panel Subsidy

Applying for a **solar panel subsidy** is a straightforward process. Here is how you can apply for the schemes:

1. PM Kusum Yojana:

- **Step 1**: Visit the official PM Kusum Yojana portal.
- **Step 2**: Register your details along with the necessary documents like Aadhaar, land ownership proof, and bank details.
- **Step 3**: Once approved, the solar system vendor will proceed with the installation.
- **Step 4**: The subsidy amount is directly transferred to the vendor, reducing your installation cost.

2. PM Solar Yojana:

- **Step 1**: Register on the **PM Surya Ghar Muft Bijli Yojana** portal with your property and system details.
- **Step 2**: Select a registered solar vendor for installation.
- **Step 3**: After installation, the vendor applies the subsidy and installs the net meter.
- **Step 4**: The system is set up, and you can start using solar power immediately.

FAQs About Solar Panel Subsidy

1. How much subsidy can I get for installing solar panels ?

The subsidy varies depending on the scheme and the system capacity. For example:

- Under the **PM Solar Yojana**, you can get up to Rs.30000/- per KW subsidy for a system equal to or under 2 kW.
- For systems 3 KW and above upto 10Kw, subsidy is Rs.60000/- +Rs.18000/-=Rs.78000/-

- For **PM Kusum Yojana**, farmers can get up to 60% subsidy on solar pump installations.

2. Can I install solar panels without a subsidy?

Yes, it is possible to install solar panels without a subsidy, but the cost will be higher. The subsidy is provided to reduce the initial installation cost, making it more affordable.

3. How long will the subsidy last?

The subsidy is available for a limited period and may change based on government policies. Be sure to check the latest notifications before applying.

4. Are there any maintenance costs after the installation?

Maintenance costs for solar systems are typically low. Vendors may offer free maintenance for the first few years. Regular cleaning and periodic inspections are recommended for optimal performance.

5. How do I apply for the solar panel subsidy ?

To apply for the subsidy, you need to:

- Register on the respective scheme's official portal (e.g., **PM Surya Ghar, PM Kusum Yojana**).
- Submit your application along with necessary documents.
- Choose an authorized solar vendor for installation.
- Once the installation is complete, the subsidy is applied to reduce the cost.

6. Can I install solar panels without a roof?

No, solar panels require a roof or open space to install. The roof should be unshaded, ideally south-facing, and strong enough to support the panels.

7. Is the subsidy available for commercial properties?

Yes, the **solar panel subsidy in India** is available for both residential and commercial properties. However, specific schemes like **PM Solar Yojana** mainly target residential installations, while others like **PM Kusum Yojana** can be used by agricultural and commercial entities.

8. What are the eligibility criteria for availing the subsidy?

Eligibility criteria for solar panel subsidy in India generally include:

- Being an Indian citizen.
- Owning the property where the solar system will be installed.
- Having a suitable roof for solar panel installation.
- In the case of farmers, proof of agricultural land is required under schemes like **PM Kusum Yojana**.

9. What is the process of net metering, and how does it work?

Net metering allows homeowners to sell excess solar power back to the grid. You generate electricity from your solar panels and use it to meet your household needs. Any surplus power is sent to the grid.

10. How much can I save with solar panels?

The savings with solar panels vary depending on your electricity consumption and system size. On average, you can save anywhere from ₹ 20,000 to ₹40,000 annually on your electricity bill by installing solar panels.

11. Are there any tax benefits associated with installing solar panels in India?

Yes, the government offers tax incentives for solar energy adoption, including deductions under the **Income Tax Act, 1961**. Investments in solar systems can be claimed as a tax deduction under Section 80-IA.Besides the solar panel subsidy in India, this tax law also benefits solar rooftop owners.

12. Can I install a solar panel system with subsidy if I live in a rented property?

Yes, you can both own a solar rooftop system and claim solar panel subsidy in India in a rented home. Although, it requires approval from your landlord. You must also ensure that the property's roof is suitable for installation.

13. What is the typical payback period for solar panels?

The payback period for solar panels in India typically ranges from 5 to 7 years, depending on your electricity consumption, the size of the system, and the solar panel subsidy you get in India. After this period, the electricity generated by the panels is essentially free.

14. Can I get a subsidy for installing solar batteries with my panels?

No. Currently, only solar panel subsidy schemes are available in India.

15. What is the cost of solar panel installation ?

The cost of installing solar panels in India can vary significantly depending on the system size, location, and type of installation. On average, the cost for a 1 kW solar system can range from ₹40,000 to ₹50,000. With solar panel subsidy offered, the cost can be reduced by up to 60% under certain circumstances.

16. Can solar panels work during power outages?

No, if your system is equipped with a **solar battery** or if you have a **solar inverter** capable of supporting power backup, your solar panels can work during power outages. This is typically more common in Hybrid or Off-grid systems with batteries.

17. Is there a limit to the amount of solar energy I can generate with the subsidy?

There is no fixed limit to the solar energy you can generate under the subsidy schemes. However, the subsidy is applicable based on the system capacity, which is usually limited to specific capacities like 2 kW or 3 kW for residential properties. For large commercial or agricultural installations, the subsidy can extend to higher capacities.

18. Can I install solar panels in remote areas or villages?

Yes, solar panels can be installed in remote areas and villages, and several government schemes like the **PM Kusum Yojana** focus specifically on rural and agricultural areas. In fact, off-grid solar systems are often preferred in remote locations with limited access to the grid.

19. How can I verify the solar vendor before installation?

Make sure the vendor has done enough work in solar and assess his technical abilities through his presentation. You can also check for certifications if any, the vendor has and past customer reviews to ensure the vendor meets quality standards.

20. How many times can I apply for Solar Panel Subsidy ?

In India, an individual can claim the solar panel subsidy only once.

The **solar panel subsidy in India** provides a significant opportunity to transition to renewable energy without the heavy initial investment. With various government schemes offering financial assistance and reducing installation costs, adopting solar energy has never been more accessible. Whether you are a farmer looking to solarize your irrigation system or a homeowner wishing to reduce your electricity bill, solar energy can be your solution.

By taking advantage of available subsidies, you can not only save on energy costs but also contribute to a greener future for India. Be sure to research the schemes, check your eligibility, and get started on your journey towards a sustainable energy solution today!

What is the Solar Subsidy for Housing Societies?

The **solar subsidy for housing societies** is a financial incentive provided by the Government of India to encourage the installation of solar rooftop systems in both individual residences and large-scale housing communities. Under the **PM Surya Ghar Muft Bijli Yojana**, housing societies can avail of subsidies to reduce the upfront cost of solar installations, making the transition to renewable energy far more affordable.

For housing societies, this scheme offers a subsidy of ₹18,000 per kW for systems up to 500 kW. This means a potential subsidy of up to ₹90 lakhs, making it a game-changer for cooperative housing societies (CHS) looking to adopt solar energy.

By utilizing this solar subsidy, housing societies can lower their electricity bills and promote sustainable living, all while contributing to India's ambitious target of achieving net-zero carbon emissions by 2070.

Why Should Housing Societies Consider Going Solar?

Switching to solar energy isn't just about savings—it's about the future. Here are several key reasons why your housing society should consider installing a solar power system:

1. Significant Electricity Bill Savings

One of the primary motivations for housing societies to adopt solar is the promise of massive electricity bill reductions. For many large residential complexes, common areas like stairwells, parking lots, elevators, and water pumps consume a considerable amount of electricity, leading to high monthly bills.

By installing a solar power system, societies can offset a significant portion of these costs, ultimately leading to **free electricity** once the initial investment is paid off. Many housing societies see a full return on their investment within **2-4 years**, after which they enjoy decades of virtually cost-free electricity.

2. Reduced Carbon Footprint

Installing a solar system helps housing societies make a positive environmental impact. Solar power generates electricity without harmful emissions, helping reduce the society's carbon footprint. This aligns with global and national initiatives toward sustainability, allowing societies to contribute to cleaner air and reduced reliance on fossil fuels.

3. Increased Property Value

With rising energy costs and growing environmental awareness, properties equipped with solar power systems are increasingly in demand.

Housing societies that invest in solar energy often experience an increase in the property's market value. Prospective buyers see solar installations as an attractive feature that will provide long-term savings and reduced reliance on grid power.

4. Energy Independence

Solar energy offers housing societies greater energy independence by reducing reliance on grid electricity. With the right size of the solar power system and net metering in place, societies can generate their own power, even feeding excess energy back into the grid to earn credits.

Key Benefits of Solar Subsidy for Housing Societies

So, how does the solar subsidy actually help housing societies? Here's a breakdown of the primary benefits:

- **Lower Installation Costs**: The **subsidy of ₹18,000 per kW** significantly reduces the total cost of the solar power system. For a system of up to 500 kW, societies can receive a subsidy of up to ₹90 lakhs, making large installations more feasible.
- **Improved Payback Period**: With the reduced upfront cost thanks to the subsidy, the payback period for solar installations in housing societies is greatly shortened. In many cases, the system pays for itself in just a few years.
- **Ease of Application**: The process to apply for the solar subsidy has been simplified for housing societies. Partnering with a reliable vendor like Prime Green Technologies Pvt Ltd makes the process even easier as we handle all the paperwork and technical requirements.

Massive Cost Savings with Solar Subsidy

Let's take an example to illustrate the cost savings:

Imagine a housing society that wants to install a **100 kW solar system**. Without the subsidy, the project cost could be approximately ₹65 lakhs. However, by applying the solar subsidy of ₹18,000 per kW, the housing society would receive **₹18 lakhs in subsidies**, bringing the total cost down to ₹47 lakhs—an impressive **28% savings**.

How to Apply for Solar Subsidy for Housing Societies

Applying for a **solar subsidy** for housing societies under the PM Surya Ghar Muft Bijli Yojana may seem daunting, but it's a straightforward process. Here's a step-by-step guide:

1. Visit the National Portal

Go to the **PM Surya Ghar Muft Bijli Yojana website** and click on 'Apply for Rooftop Solar.' This will redirect you to the application form where you need to fill in the required details.

2. Submit Registration Information

Enter your state, district, and DISCOM (electricity distribution company) details. This is followed by providing your customer account number and basic personal information.

3. Provide Necessary Documents

During the application process, you'll need to upload several documents such as proof of identity, proof of address, and society registration certificates.

4. Feasibility Clearance

Once the application is submitted, the local DISCOM will assess your property for the feasibility of a solar installation in your housing society. This includes a review of the available roof space and structural integrity.

5. Installation and Commissioning

After feasibility clearance, the housing society can move forward with installing the solar panels. Once installed, the system is tested and approved by the DISCOM, who will issue a commissioning certificate.

6. Receive Subsidy

Upon successful installation and commissioning, the society's bank account will be credited with the **subsidy amount** within a few weeks.

Documents Required for Solar Subsidy Application

To successfully apply for the **solar subsidy**, housing societies will need to gather the following documents:

- **Proof of Identity**: Aadhaar card or PAN card of the society's authorised signatory.
- **Proof of Address**: Utility bill or registration documents of the society's registered office.
- **Society Registration Certificate**: Proof of the society's legal existence.
- **Electricity Bills**: Recent electricity bills for common areas where the solar system will be installed.
- **Roof Ownership Certificate**: Proof of ownership or legal right to use the roof for solar panel installation.
- **Feasibility Report**: A technical report from your DISCOM or solar vendor verifying the suitability of the property for solar installation.

Partnering with a Trusted Solar Vendor

One of the most important aspects of applying for the solar subsidy for housing societies is choosing a reliable solar vendor. At **Prime Green Technologies Pvt Lts**, we guide housing societies through every stage of the process, from initial consultation to final installation and post-installation support.

- **Expertise in Solar Installations**: With years of experience in installing solar systems for both individual homes and housing societies, Prime Green Technologies Pvt Ltd ensures your solar installation is completed with precision and efficiency.
- **Subsidy Application Assistance**: We handle the entire subsidy application process on your behalf, ensuring that you secure the full benefits of the **PM Surya Ghar Muft Bijli Yojana.**

Looking for a solar rooftop for housing societies then you must connect with us. Drop us a text on WhatsApp or call +91-8104414765 (Manoj Kalke) to know more.

Frequently Asked Questions (FAQs)

1. How much subsidy can housing societies get under the PM Surya Ghar Muft Bijli Yojana?

Housing societies can avail of   **₹18,000 per kW**, up to a maximum of **₹90 lakhs** for installations up to 500 kW.

2. What is the payback period for solar installations in housing societies?

With the subsidy, most housing societies recover their solar investment within **2-4 years**, after which they benefit from free electricity.

3. Can a solar vendor assist with the subsidy application process?

Yes, partnering with a reputable solar vendor like **Prime Green Technologies Pvt Ltd** can significantly simplify the application process and ensure that you receive the subsidy efficiently.

By choosing solar, housing societies can unlock incredible long-term savings, reduce their carbon footprint, and contribute to India's green energy future. With the **PM Surya Ghar Muft Bijli Yojana** offering substantial subsidies, there has never been a better time for your housing society to make the switch.

CHAPTER THIRTEEN

REFERENCES

1.

From Sunlight to Electricity : A practical handbook on solar photovoltaic applications (Third Edition) by TERI, India

1.

Solar Photovoltaics (Third Edition) by Prof Chetan Singh Solanki

3.

Ministry of New & Renewable Energy website www.mnre.gov.in

4.

Training Manual for Engineers on Solar PV system, Technical Report of Year 2011, Tribuvan University, Nepal

5.

Energy Statistics Manual by IEA, USA

6.

Solar Electric System Design, Operation & Installation- an overview for builders in the US pacific North West, Oct 2009.